ISAAC NEWTON

ISAAC NEWTON
THE ASSHOLE WHO
REINVENTED THE UNIVERSE

Florian Freistetter
Translated by Brian Taylor

 Prometheus Books

59 John Glenn Drive
Amherst, New York 14228

Published 2018 by Prometheus Books

Originally published by Carl Hanser Verlag GmbH & Co. KG, München, in March 2017 as *NEWTON: Wie Ein Arschloch Das Universum Neu Erfand*.

The internet addresses listed in the text were accurate at the time of publication. The inclusion of a website does not indicate an endorsement by the author(s) or by Prometheus Books, and Prometheus Books does not guarantee the accuracy of the information presented at these sites.

Cover design by Jacqueline Nasso Cooke
Cover and interior illustrations by Büro Alba
Cover design © Prometheus Books

Inquiries should be addressed to
Prometheus Books
59 John Glenn Drive
Amherst, New York 14228
VOICE: 716–691–0133 • FAX: 716–691–0137
WWW.PROMETHEUSBOOKS.COM

22 21 20 19 18   5 4 3 2 1

Library of Congress Cataloging-in-Publication Data

Identifiers: ISBN 9781633884571 (ebook) |
ISBN 9781633884564 (hardcover)

Printed in the United States of America

CONTENTS

Newton was able to combine mental power so extraordinary that if I were speaking fifty years ago, as I am old enough to have done, I should have said that his was the greatest mind that any man had ever been endowed with. And he contrived to combine the exercise of that wonderful mind with credulity, with superstition, with delusion, which would not have imposed on a moderately intelligent rabbit.

—George Bernard Shaw

INTRODUCTION

Isaac Newton was born on Christmas Day in the year 1642. Or on January 4 in the year 1643. It depends entirely on which calendar you consult—and the fact that there were two different calendars in use back then is evidence enough of the confused times in which the man who would later reinvent the universe was born.[1] It was high time for a genius to appear and shed light on the world, and Isaac Newton was that very genius. And an asshole to boot.

People generally think of Newton as the man who almost singlehandedly founded the modern physical sciences, as the scientist whose work forms the basis of practically all branches of the natural sciences, as the person who gave us a completely new perspective of the universe. All of that is absolutely true. And yet Isaac Newton was also a selfish and belligerent bully. I don't say that because I have anything against the man. On the contrary—my own work as an astronomer is dominated by Newton's achievements. Without Newton, none of the things I have found out in the course of my astronomic research would have been possible. I worship Newton more than almost any other scientist of the past—even though he was such a jerk.

In fact, to a small degree, I admire him precisely because he was so horrible. Does that sound strange to you? A contradiction in terms? It's to do with the fact that, in the modern world of science, the most important thing is to stick to the social and political rules of the research apparatchiks and to behave in an opportune fashion. Otherwise, you'll never have a successful career. So there simply couldn't be somebody like Newton today—a person who seems to delight in making enemies his whole life long and yet still manages to revolutionize the world of science. Or perhaps there could. Can a genius and asshole be a role model for a successful scientist even today?

Isaac Newton was the guy with the apple and gravity, as most people probably know. But his achievements went much further than that. There was practically no subject that he left untouched, so it's worth taking a closer look at his work. But the same is true of his fascinating character: he was odd, unfriendly, uncompromising, extremely sure of himself, resentful, argumentative, secretive, insensitive, underhanded, and a religious fanatic who liked to prophesy the end of the world. But—and this is the most interesting point—if he hadn't been all of that, he most likely wouldn't have been in a position to change the world in the way he did.

Newton is a man from another time. His scientific achievements, however, have survived up until the present day and will continue to be valid in the future. The natural world can still be explained using the laws that he established back then. In this book, which makes no claim to be a comprehensive biography—if we wanted to understand Newton's life and work as fully as is possible some three hundred years after his death, we would have to write entire libraries of books, which experts in the past have indeed done (see, for example, the recommended reading at the end of this book)[2]—I take a look at the way in which Newton treated himself and his fellow human beings, complete with all the disputes, absurd decisions, and bizarre episodes, which find no mention in the usual physics text books.

Newton was a strange and awkward chap from the very beginning. "What shall become of me? I will make an end of it. I can only weep. I do not know what to do."[3] He wrote these depressing lines in his notebook while still a youth. After a not particularly happy childhood, he saw no prospect of a happier life as an adult. His father had died before his birth, and his mother remarried when Isaac was three years old. The child seemed to be viewed as a nuisance for the young couple and was sent off to live with his grandmother. At school, he was cleverer by some distance than his classmates, which only served to heighten his sense of isolation.

A list of all Newton's "sins," as written later, at the age

of nineteen, in his notebook, gives an indication of his relationship with his family and other people in general. When he writes that he was guilty of "peevishness with my mother," that's understandable. Entries like "Stealing cherry cobs from Eduard Storer" or "Calling Dorothy Rose a jade" also sound like things that youngsters are prone to do, just as pretty much every boy has at some point had "unclean thoughts words and actions and dreams." "Making pies on Sunday night" or "Making a mousetrap on Sunday" are no longer considered misdeeds today,[4] though they obviously lay heavy on Newton's heart. But some entries provide a deeper insight into his character: "Wishing death and hoping it to some" or "Striking many." And by the time we read "Threatening my father and mother Smith to burn them and the house over them," it is clear that Isaac Newton's head was full of thoughts that have no place in a normal childhood and youth.

It follows that Newton preferred to focus on the physical world around him than on other people. He was fascinated by machines, constructed water mills, and observed how, when light passed through tiny holes in the wall of his attic bedroom, the sun's course could be measured with the rays of light. He then made sundials that were so accurate that the family he lived with during his school days used them to tell the time. When Isaac was sixteen, his mother—now a widow once more—summoned him back home to take over the family's farming activities, which ran completely contrary to his interests.

Farming was not one of Newton's fortes. He might have been better at it if he had had any interest at all in the work. But even then, he was above all a difficult, selfish, and childish young man—one who had no desire for tough physical activity. Fortunately for him and for science, however, his uncle and his former teacher were able to convince Isaac's mother to allow him to attend the University of Cambridge.

In June 1661, Newton left his home village of Woolsthorpe in the county of Lincolnshire and set off for the university. There, he began the work that would shake the foundations of the world as it was then and whose effects can still be felt today. There, he also began the next stage of the development of his character, such that we can say in hindsight, "What a scientific genius! And what an asshole!"

CHAPTER 1 AT ALL COSTS: NEWTON, RUTHLESS IN THE EXTREME

ven today, scientists are still often considered to be an odd bunch—socially isolated, badly dressed, incapable of talking about anything other than their work, and fully focused on research, they have no private life and spend their whole time locked away in their labs. And even if that is a complete exaggeration, it would fit Isaac Newton very well. He was the epitome of a nerd. In itself, that's a likeable thing—and the nerdiness of the seventeenth century will come across in this chapter as a rather sweet characteristic. But Newton took his uncompromising nature to the extreme. When he was trying to find out something about the world, he had no regard for his own sensitivities and those of other people, or for social conventions. His thirst for knowledge resulted in a radical approach, which caused him not infrequently to cross the line into asshole territory.

Humphrey Newton, Isaac's assistant in Cambridge (but no relation), wrote of Newton's behavior: "I never knew him take any Recreation or Pastime. . . . Thinking all Hours lost, that was not spent in his Studyes . . . So intent, so serious upon his Studies, that he eat very sparingly, nay, oftimes he has forget to eat at all, so that going into his Chamber, I have found his Mess untouch'd, of which when I have reminded him, would reply, Have I; & then making to the Table, would eat a bit or two standing, for I cannot say, I ever saw Him sit at Table by himself . . . He very rarely went to Bed, till 2 or 3 of the clock, sometimes not till 5 or 6, lying about 4 or 5 hours . . . He very rarely went to Dine in the Hall unless upon some Publick Dayes, & then if He has not been minded, would go very carelessly, with shoes down at Heels, stockins unty'd, surplice on, & his Head scarcely comb'd."[1]

Even when Newton did take time for a quick stroll, it would often occur that, inspired by a sudden thought, he would hasten back to his study and get back to work at his desk—standing up, as Humphrey Newton describes, since Isaac's time was obviously too precious to be wasted on the act of sitting down.

Sadly, no records have been left by students who attended his lectures. This may have something to do with the fact that their number was extremely limited: "when he read in the Schools . . . where so few went to hear Him, & fewer that understood him," writes Humphrey Newton.[2] Isaac Newton seems not to have minded this at all; if the students stayed away from his lectures, he would simply speak to the walls of the empty hall.

Apart from the world in his head, the only other thing that seems to have been of interest to Newton during his time at Cambridge was his vegetable garden, where he would go for walks and regularly pulled out the weeds that, according to Humphrey Newton, he simply couldn't stand.

Little sleep, eating on his feet, lectures to empty halls, and a distaste for weeds—right from the beginning, Isaac Newton ticked all the boxes as a mad professor. But he was far from being a harmless figure of fun; when it came to understanding the world, his rigor knew no bounds.

"Fresh air, fasting and little wine," was the recipe for a successful scientific career that he jotted down in his notebook. But he also wrote that "too much study (whence . . . cometh madnesse)."[3] Perhaps he should have paid more attention to his own advice, since when we look at the way he worked, it would appear as though he had long since crossed the line to insanity.

Take the time when, as a young man, he stuck a needle into his eye in order to find out more about the nature of light. A completely crazy thing to do, but he wanted to gain knowledge at all costs. As already mentioned, Newton's rigor knew no bounds—and he was definitely not a softie.

TALKING DOGS AND VOMITING SALAMANDERS

Let's not forget the time in which Newton lived, though. At the end of the seventeenth century, there were no "natural sciences" in the modern sense of the term. Nature was full of mystery, and unanswered questions abounded about almost every aspect of it. Newton wasn't alone in his peculiarity—the entire world of science he lived in seems absurd to us now. You only need to take a look at the publications of the time. In March 1665, the first edition of the *Philosophical Transactions of the Royal Society* appeared, in which scholars of the time discussed their findings. The *Philosophical Transactions*, along with the French *Journal des sçavans*, founded three months earlier, are the oldest scientific publications in the world.

Though, one might doubt their scientific value when one looks at the articles published therein. In an early edition, for instance, a Mr. Colepresse reports on "A relation of an uncommon accident in two aged persons": Joseph Shute, eighty-one, and Maria Stert, a mother of nine aged seventy-five, both grew new teeth, despite their advanced years. And this occurrence was obviously considered to be remarkable enough to be published in the new scientific journal!

But that's nothing compared to "An extract of a letter not long since written from Rome, rectifying the relation of salamanders living in fire."[4] A "gentleman," whose name is not given, had cast a salamander from overseas into a fire in order to see what would happen. The creature began vomiting to extinguish the flames and continued to do so until it was removed from the fire two hours later in order not to "hazard" it any further, as the report puts it. The veracity of this story is doubtful, as is the case with many other "research reports" of the time. Everything that was in any way unusual was considered to be worth reporting—a stone that was found in the head of a serpent, the birth of a calf with two heads, an English merchant bitten by a snake in Syria, a mysterious shower of fish that fell on England (recorded by an "honourable gentleman" who gathered these curious fish that had fallen from the heavens and preserved them, only to misplace them, alas, and thus not be able to present them for inspection).

The random and untamed curiosity of the scientists of the time is perhaps best illustrated by an article that appeared in the fourth edition of the *Philosophical Transactions* with the title "An extract of M. Dela Quintiny's Letter, written to the publisher in French sometime agoe, concerning his way of ordering melons; now communicated in English for the satisfaction of several curious melonists in England."[5] An incredible four whole pages are devoted to the description of how melons should be arranged in beds, how their leaves should be trimmed, and how they should be treated and harvested, complete with corresponding diagrams.

The world was one big mystery, and people were busy trying to decipher it. They didn't call themselves "scientists" at the time. Those who, like Newton, were busy trying to understand the world using scientific methods were "natural philosophers." What they were doing did indeed include traces of what we understand today as philosophy, but ultimately it would over the course of time become true natural science. For that to happen, it was necessary to finally begin using new methods to try to understand the world, and melons, deformed baby cows, or vomiting, fire-extinguishing salamanders were just as interesting in this regard as what we would call "true science" today. Renowned scientists like Robert Hooke or Robert Boyle wrote in the *Philosophical Transactions* about their work with vacuum pumps, astronomic observations, or new optical devices. Boyle, for instance, laid the foundations for modern chemistry back then and developed what is known as Boyle's Law (or the Boyle-Mariotte Law), about the characteristics of an ideal gas,[6] a law that is still taught at every university today. But he found nothing unusual about also publishing an article called "Observations upon a monstrous Head of a Colt," in which he went into detail about the deformed skull of a newborn foal. Gottfried Wilhelm Leibniz, one of the greatest scholars of the time (and one of Newton's archenemies—see chapter 7) is known today with some justification as a mathematical genius, an eminent philosopher, and a computer pioneer, and is often referred to as the last polymath. And yet he also wrote articles for the *Journal des sçavans* containing reports of a goat with an extremely unusual haircut. Apparently, this animal lived in Zwickau at the home of a Herr Winckel and had at first completely normal hair, before developing a strange hairstyle after it kicked a passerby and was therefore locked away. Leibniz, every inch the curious natural scientist, wondered why this should be and speculated that it was probably the goat's mourning for the loss of its freedom that was responsible for the change in its coat and hair.

Some years later, Leibniz wrote about a talking dog in the town of Zeitz in Saxony-Anhalt. A child had taught the dog a few words, such as *thé* ("tea"), *caffé* ("coffee"), or *chocolat* ("chocolate").[7] Leibniz's reputation didn't suffer in the least as a result of these reports, since they were considered perfectly normal by seventeenth-century standards. When practically nothing is known about the world, dogs that can apparently talk are almost as useful as sources of information as astronomic observations are. And even what we now consider to be "serious" science didn't follow the same procedures as those we are used to today. When Robert Hooke (another major scientific figure who would later count among Newton's enemies—see chapters 3 and 4) wanted to understand how the breathing system worked, he simply grabbed a dog, cut the living creature open, and blew air into its lungs with a pair of bellows. The dog survived (though only for a short time, alas), and Hooke began to wonder if he could perhaps learn more if he were to divert the blood from the veins and bring it into contact with fresh air.

RESEARCH, REGARDLESS OF THE CONSEQUENCES

In this strange world of burning salamanders, it was a considerable feat to be even more strange and curious, but Isaac Newton managed this without any difficulty at all. It wasn't enough for him simply to trust what some "honourable gentleman" had reported. He had no interest in the scholastic traditions of the universities, where it was still common practice to interpret the texts of the ancient Greeks. He had no time for hypotheses and only wanted reliable knowledge gained from actual experiments. He wanted to find out things for himself, and from bottom to top—a completely new approach at the time, and one that met with incomprehension.

Even as a schoolboy, Newton filled his notebook with information and questions of all kinds. He copied out texts from other books, for instance instructions on how to draw landscapes, melt metal, or catch birds (apparently, it works best if you make them drunk with wine). He was fascinated by light and color even before he carried out his revolutionary research on them, and he put together a list of the different paint colors that he knew of. His view of the relative importance of these is slightly bizarre. "A colour for dead corpses" and "Colours for naked pictures" have their own section, whereas the rest of the list contains rather lackluster descriptions, such as "A green," "Another green" or "A light green." But then, Isaac was still a teenager at the time, and nudes and corpses were of course of greater interest.

We can also find in his notebook a list of things which are painful for the eye. Dust, fire, and "too many tears" sound reasonable enough, as do "garlic," "leek," and "onions." But why Newton considered "warm wine" to be detrimental to the eyes remains a mystery—perhaps only disgusting plonk could be had in England at the time. "Sticking a needle into your eye" is not on the list, but that is exactly what Newton did a few years later.

He didn't do it just for fun—even Newton wasn't that crazy. The aim of this experiment was to solve an important question: How do our sense organs work? This was another thing about which little was known at the time. Ancient Greek philosophers like Euclid or Ptolemy thought that the eye emitted mysterious rays that rendered objects visible. Aristotle and his followers, on the other hand, believed that the objects themselves projected rays into our eyes and could thus be seen. People hadn't gotten much further than this in the seventeenth century, and this bothered Newton, who asked himself the following: If we don't know how our sense organs work, but it is these senses that enable us to perceive the world, then how the hell are we supposed to understand the world?

A fair question, and perhaps a good enough reason to tinker around with your own eye with a big fat needle. Newton stuck the thing between his eyeball and the socket in such a way that he could press the eyeball from behind with the tip of the needle. He was thus able to change the shape of the eyeball and observe the effect this had on his visual perception. He saw rings of differing sizes and colors that changed when he moved the needle, but disappeared when it was still, as though light was something that arose through pressure. An interesting observation indeed, since as little was known at the time about the exact nature of light as about the way the eye worked.

Newton wasn't satisfied with sticking a needle in his eye, however. These days, when there is a solar eclipse, an astronomer can hardly mention it without having to give repeated warnings not, in any circumstances, to look directly at the sun with the naked eye, since this is incredibly dangerous (which unfailingly leads to everybody suddenly wanting to buy eclipse glasses, so that the shops run out and people then use other, potentially risky, means to protect their eyes).[8] So what did Newton do? Precisely that—he simply stared at the sun with his naked eye for as long as he could, just to see what would happen. He survived all of these self-experiments without losing his sight and was deeply fascinated by the effects that arose. If he stared at the sun for long enough, he could later still see afterimages and curious colors that were obviously not real. Despite the interesting fruits of his research, Newton was actually slightly concerned about his eyes and locked himself away in a completely darkened room for three days, only reemerging when they were working normally again.

Just as Newton, in his thirst for knowledge, had little regard for himself and his own health, he was equally inconsiderate of the sensitivities of others. If they even dared to cast doubt upon the things he had painstakingly found out for himself, he showed no mercy or even understanding in his reactions, as many of his contemporaries discovered (in particular his counterpart Robert Hooke, see chapters 3 and 4).

This is somehow understandable, too. After all, if you go to such great lengths as he did, then it's normal to be a bit sensitive if others don't appreciate your efforts. If you stick a needle into your eye, it's not difficult to feel a bit pissed off when people then go on to criticize you for it. Especially if your critics have no idea what they are talking about. And Newton was absolutely convinced that his critics didn't have the faintest idea. Nevertheless, a little understanding on his part would have been advisable. His scientific methods were certainly out of the ordinary for most of his contemporaries, and he certainly took a different approach from scholars before him. But he wasn't completely removed from his time; he was familiar with the world of the universities and their customs. He must have been aware that much value was placed on academic debate about philosophical ideas and that his refusal to participate in it was bound to meet with incomprehension.

But Newton was not an understanding man. He couldn't see any point in taking into account the views of those around him. He was completely focused on his own view of the world and was offended that nobody had any inclination to appreciate his new methods of investigation. Newton carried out experiments to test out hypotheses. In other words, he did exactly what natural scientists today always do as a matter of course. But in the seventeenth century, many scientists still had to get used to this idea.

Newton's ruthless approach in pursuing his goals can be seen not only at the beginning of his career but also to an equal extent at the end. He lived until the age of eighty-four—quite an achievement in those days—but made practically all of his major scientific discoveries during the first thirty years of his life. For the last thirty years, he busied himself in a completely different field—and was as unrelenting in this as ever before.

THE WARDEN AND THE FORGER

So let's jump from the young Newton to the old one. Much had changed (and the things that Newton had experienced in the meantime are the subject of the following chapters); much, though, remained the same. Instead of arguing with his peers as an unknown scientist and ordinary member of the Royal Society, Newton now argued with them as a famous scientist and president of the Royal Society. And he was as uncompromising as in his youth. Probably even more so. For, in his old age, Newton got to live out, at least to an extent, every nerd's secret dream: he left the university and began to fight crime. Not as an early superhero complete with a cloak and supernatural powers,[9] but he was at least armed with an impressive title: the warden of the Royal Mint.

Newton held this post from 1696 onward. Until that year, he had scarcely ventured out of his customary surroundings, his life being split between his home village of Woolsthorpe and the leisurely tranquility of academia in Cambridge, some 100 km away. He paid his first visit to London in 1668 and seldom returned in the years that followed. It was only later that he became a regular visitor to the capital city, above all in 1689 and 1690, when he held a seat in the English Parliament. Little is known about these early stages of Newton's political career. He seems not to have given any stirring speeches or otherwise distinguished himself. The only evidence of activity on his part concerned a complaint about a cold draught.

He seems to have liked it in the big city, however. In the years following his spell in Parliament, the isolation of life in Cambridge must have been too dull for him, and he started looking for new challenges in London.[10] He needed to find a suitable job there to earn a living and was fortunate enough to have a niece, Catherine Barton, who was involved with Charles Montagu, the Earl of Halifax and the queen's chancellor. Newton obviously took advantage of this relationship in order to land a well-paid post at the Royal Mint.[11]

Others might well have taken it easy after obtaining such an administrative post and spent their twilight years enjoying peaceful prosperity at the state's expense. Not Newton, however, who continued to demonstrate the same pigheadedness that had been his trademark during his scientific career. He needed to as well, since he was confronted by not only a dilapidated financial system, but also an opponent of a completely different caliber from the scholars with whom he had till now crossed swords.

Newton was no superhero during his crime-fighting years—similarly, his adversary was anything but a classic super-villain. Had William Chaloner chosen anyone other than Isaac Newton as his opponent, it is likely that his name would have fallen into obscurity today. We do not know his exact date of birth and his origins were unremarkable. He trained as a smith in Birmingham, and didn't even do that properly, learning only the specialist trade of nail-maker, a profession that was already on the verge of dying out. At the end of the seventeenth century, machines were first used in the production of nails, and nail-makers like Chaloner were consigned to being just poorly paid laborers. It is no wonder that many metalworkers of the time turned away from nails to the more lucrative activity of forging money. If you have to spend your days working on small pieces of metal, you might just as well make coins, Birmingham's smiths thought to themselves, and put this idea into action with such enthusiasm that the forged coins became known as "Birmingham groats"[12] and were for a time more commonly available than the genuine version.

It is not recorded whether Chaloner was already producing forged money at the time, but he certainly had the requisite skills to do so. When he left Birmingham to try his luck in London, he began by applying his talents in another direction. Contemporary sources are rather vague when it comes to describing the product that Chaloner dealt in: "The first part of his Ingenuity shoed it self in making Tin Watches, with D-does &c in 'em.'" What Chaloner's biography *Guzman Redivivus* somewhat bashfully calls "D-does &c" seems to have been no more than a sex toy.[13] Chaloner obviously made a kind of pocket watch with inbuilt dildos. Highly sophisticated stuff.

His early criminal activities were equally illustrious. He conned people into buying pseudo-medical tinctures and acted as a soothsayer. He was particularly skilled at predicting where people would find possessions that had been lost or stolen, an ability that was no doubt linked to the fact that he himself had pinched the things in the first place. He also learned japanning and gilding on the side, an art that would later help him with his coin forging. And there was certainly opportunity enough to exploit the financial system in Great Britain toward the end of the seventeenth century.

Britain's currency was in rather dire straits. The old silver coins had long since ceased to be worth their supposed value. It was easy enough to cut a tiny bit off the edge and collect the silver thus gathered. This practice, which was called "clipping," reduced the weight and thus the value of the coins, leaving a nice profit for the "clippers." The Royal Mint tried to undermine the practice by issuing new coins that were also embossed on the edges, meaning that any clipping would be easily noticed. Another problem remained, however: the silver used for the coins was worth more on the European continent than the coins' nominal value in Britain. Traders hoarded the English coins, therefore, and shipped them to Paris or Amsterdam, where they could be melted down into silver bars and sold for a profit.

England's coinage was in the process of disappearing. The coins that remained were constantly losing value because of clipping—or were indeed counterfeit. It is estimated that 10 percent of all coins at the time were not produced by the Royal Mint and were instead made by counterfeiters like William Chaloner.

Isaac Newton wasted no time in his new job. His first suggestion was for a completely new coinage for Britain's currency. All the old coins were to be withdrawn from circulation and replaced by new, more secure ones. In order to execute this idea as quickly as possible, he needed to put his talents to use in a wholly new fashion. He studied the production methods, examined the machines, and observed the workers. He analyzed each and every process involved, measured the time taken, and worked out how to optimize them all. He calculated the ideal frequency at which hammers should be hammered, presses should be pressed, and workers should work, and it was no surprise that he was extremely successful in all this. While the Mint could previously produce just about 15,000 pounds' worth of new coins per week, Newton's improvements led to this figure rising to 50,000 pounds per week (with highs of up to 100,000 pounds per week).[14]

After a surprisingly short time, the replacement of the coinage had been completed, but the counterfeiting problem was still there. Chaloner and the many like him remained active. However, while most of the criminals were content to stick to forging money, Chaloner had set his sights higher, namely, on the place where the most profit was to be made— on the Royal Mint itself.

To this end, he promoted himself in the public eye as an expert who could solve the problems besetting the financial system. He composed pamphlets with titles like "Proposals Humbly Offered, for Passing an Act to Prevent Clipping and Counterfeiting of Money" and did indeed manage to attract the attention of several influential figures, to whom he made suggestions for new machines and processes that would enable the production of supposedly counterfeit-proof coins. The only thing he required was access to the Mint, so that he could carry out his experiments there. At the same time, he took every opportunity to discredit those already working at the Mint and accused them of criminal activities themselves.

In short, Chaloner did not merely claim to know better than Newton how to run the Royal Mint; on top of that, he accused him of being involved in dodgy dealings. Chaloner need not have brought out such heavy artillery to turn such an easily offended man as Newton against him, but by so doing, he managed with all flags flying to make a sworn enemy out of him.

CUT TO THE QUICK

Once Newton had completed the replacement of the coinage, he devoted himself completely to another of his duties as warden: the pursuit of counterfeiters. He did so with the same wholehearted persistence that he applied to everything else in his life, as extracts from his expense accounts demonstrate. He created a network of informers throughout London and equipped them with suitable clothing so that they would not stand out in the world of petty criminals. He himself frequented taverns and jails in order to question suspects and those in custody. He had many people locked up in cells at the Royal Mint in order to be able to interrogate them at his leisure. We do not know how he went about this, but it is unlikely that he was particularly friendly. The written records of the interrogations have mostly disappeared—many of them were burnt by Newton himself or otherwise destroyed on his orders. Whether he simply wanted to get rid of old paper, or didn't want posterity to hear about the interrogation methods he used, we do not know.

For all Newton's dedication to law and order, one man managed continually to evade him: William Chaloner did keep on ending up in prison, but he never remained there for long. He often played a similar game to the one he had practiced as a soothsayer: he would forge money (including the newly introduced banknotes) and, when the matter was exposed, would explain to the banks how the money had been forged and what they could do in order to prevent this in the future. The information not only bought his freedom—it often got him a reward as well.

His main target remained the Royal Mint, however. In 1697, he did indeed receive permission to carry out his experiments there with new machines. Newton himself was supposed to prepare everything and show Chaloner all the things that outsiders would normally never get to see. But he simply refused to do so—and ended up getting his way. The Royal Mint remained closed to Chaloner, though a few of the allegations he made about Newton refused to go away, meaning that the latter had no choice but to put up with the criticism (not an easy thing for him, see chapter 3).

At least to begin with, for he soon had the chance to deal with his enemy once and for all. In 1698, Chaloner was in jail once again (this time for forging lottery tickets), and this time Newton did his utmost to gather enough evidence to make a conviction inevitable. He prepared for the trial as methodically as he would have for a scientific treatise, attempting to obtain proof of every single one of Chaloner's offenses. And his aim was for Chaloner to die on the gallows, should he be convicted.

He questioned witnesses, once even for ten days in a row, and gathered statement after statement. In the meantime, Chaloner began to get nervous and wrote letters to Newton from his cell, attempting to talk his way out of the situation.

Yet this technique, which had always worked so well with others, failed with Newton. Chaloner's first letter was still rather confident: "I am not guilty of any Crime," he claimed therein. Newton gave no response. Chaloner then became more dramatic: "I have been guilty of no Crime these 6 years . . . if I die I am murdered." Newton gave no response. The verdict in Chaloner's trial was returned on 14 March 1699. He was found guilty. He wrote a last desperate letter to Newton: "Nobody can save me but you. O God my God I shall be murdered unless you save me. O I hope God will move your heart with mercy pitty to do this thing for me. I am Your near murdered humble Servant." Again, there was no response from Newton, and on April 1, 1699, William Chaloner was hanged.[15]

IN PRAISE OF MODERATE RUTHLESSNESS

Isaac Newton was an uncompromising and pigheaded man, going far beyond the clichéd image of the obdurate, mad professor. Without such characteristics, he would probably have never become such a great scientist. If you want to be successful in the world of science today, it's okay to be a bit odd. But it's by no means a prerequisite—whatever popular TV series like *The Big Bang Theory* may suggest. It has long since ceased to be true that all scientists are as nerdish and detached from reality as we like to imagine. Having said that, such clichés don't come from nowhere. One of the reasons why they hold true to an extent is the work itself. In modern science, you deal with highly specialized subjects that are scarcely comprehensible to the rest of humanity and it is hardly surprising that a certain amount of social isolation can arise. On top of that, universities and research institutions offer a kind of "protected environment" for people who are, or want to be, somewhat different. Unlike in many professions, it (generally) doesn't matter one bit how you are dressed, for example, as long as your work is good. Regular working hours don't necessarily need to be kept to, either—as long as suitable, scientifically useful results are achieved, it rarely bothers anybody else if you prefer to work through the night and sleep all day.

A nerd like Isaac Newton, with all his quirks and foibles and completely focused on his work, would probably have no difficulty fitting into the world of science today. It would more likely be his inconsiderateness that would lead to difficulties. Though having said that, he could also set an example in this regard, too, at least if we take a pragmatic view of things: inconsiderateness is most certainly not a positive quality, but sadly it is one that is needed far too often in the modern world of science.

Natural sciences in the seventeenth century promised neither a secure job nor an important career (something they still don't guarantee today). They were often a pastime for well-to-do gentlemen—who, because of their prosperity, could treat others badly, as Isaac Newton did, with impunity. Today, the life of a natural scientist is dominated by sponsorship, project proposals, research grants, raising external funds,[16] and job applications. If you can't promote yourself or your work, you don't have much chance. Vacant positions at universities are rare and those who make too many enemies are generally left empty-handed. If you want funding for research, you have to stick to strict guidelines and can't simply do what you feel like. But a certain uncompromising approach to your colleagues is needed.

When natural sciences were still novel enough that you could write papers stretching to pages about goats' hairstyles or the correct ordering of melons, it was easier to be a lone wolf. People knew very little about the world and there was plenty that could be researched. Nowadays, the natural sciences have become so specialized that it's practically impossible to make any progress alone. In almost every discipline, the important breakthroughs are now only made within the framework of major international collaborations involving hundreds or thousands of people. Discoveries like that of the Higgs boson at the European Organization for Nuclear Research (CERN), the detection of gravitational waves at the LIGO observatory, or the mapping of cosmic background radiation with the Planck satellite could never have been accomplished by uncooperative mavericks like Newton.

And yet: a tiny bit of inconsiderateness (or perhaps it would be better to say courage and self-confidence) wouldn't do young scientists today any harm. Even a team needs people who are prepared to go against the majority opinion if necessary. Natural sciences aren't democratic and the majority isn't always right. If you don't dare to present your own opinion, you might miss the chance for a major breakthrough. Universities are becoming more and more like a regimented system for education, where quickly gaining your degree is the main thing, not developing independent thought. Even those with PhDs all too often merely accept assignments from their professors, rather than carrying out independent research, as they really should. If you don't quickly learn to stand up for your own opinions, you'll never have the chance to tread a scientific path and come up with your own scientific results.

You can and should, therefore, be a little inconsiderate of others, if you want to forge ahead in the world of science. You don't need to go over the top like Isaac Newton did. For all his genius, he would probably be mainly viewed today as a troublemaker.

When we consider the current working conditions for young scientists, we can see that it's probably best to take the older Isaac Newton as a role model. A well-paid official post in a financial administration, and one that allows you to print your own money—definitely not a bad career move. . . .

CHAPTER 2

PRINCIPIA FIRST: NEWTON THE EGOIST

John Flamsteed actually just wanted to observe the heavens. It was his goal to put together a catalogue of the stars that was better and more comprehensive than anything that had been produced before. He did not live to see the publication of his life's work, however, and one of the people responsible for this was Isaac Newton.

"Newton very ungratefully set himself to hinder the work by several tricks and artifices . . . gave me all the trouble he could," Flamsteed wrote to a friend in October 1715, adding in despair, "how unworthily, nay, treacherously, I am dealt with by Sir I. Newton."[1] And yet the relationship between Newton and Flamsteed had actually gotten off to a good start. But Newton's ego meant that conflict was unavoidable in the end. The great scientist was too sure of himself and his abilities and could simply not admit to having made mistakes. And the one who had to suffer was a man who wasn't actually responsible for the dispute and who had indeed—at least in the beginning—wanted to help Newton.

Isaac Newton's monumental work *Philosophiae Naturalis Principia Mathematica* laid the foundation for modern natural sciences (see chapter 5). Published in 1687, the book not only contained a mathematical analysis of gravitational force, but was also in principle a completely new explanation of the world. Above all, Newton demonstrated in the book that the same laws apply here on Earth as in the heavens—that projectiles such as cannon balls follow the same rules as planets, comets, and other celestial bodies. In order to be able to demonstrate this, however, he required concrete data and observations, and that's where John Flamsteed came in, a man who spent each night studying the sky from Greenwich Observatory. But though the working relationship between the astronomer and the physicist started off on a relatively harmonious basis, it later developed into a bitter feud, sparked and fanned by Newton's inconsiderate behavior.

In March 1675, a royal decree declared Flamsteed "the King's Astronomical Observator," making him the first "Astronomer Royal." Today, this is an honorary title, held by a completely normal scientist with no special duties attached. Flamsteed's responsibilities, on the other hand, were clearly defined: he was supposed to observe the stars in the heavens and chart their positions as accurately as possible, in order to provide a solution to the problem of determining longitude at sea.

From today's perspective, it seems odd to appoint an astronomer especially for such a particular project. The English king didn't stop there, however—he even founded a separate observatory for the purpose:[2] a lot of effort for something that any smartphone today can do in a few seconds. Wherever you are in the world, your phone's GPS function will tell you your exact location and give the longitude and latitude to within a few meters. But in the seventeenth century, there were neither smartphones nor GPS—just loads of ships that were unable to determine their location, which meant that they got lost on the oceans and their crews died of hunger or drowned when the ships sank. For a country whose military and economic power was as dependent on safe and effective seafaring as England, this state of affairs was simply intolerable.

LOOKING FOR LONGITUDE

Determining latitude was no longer really a problem at the time. Latitude is the measurement of how far north or south of the equator we are. To find it out, it's sufficient just to observe the sun, for example. The sun rises in the east, sets in the west and reaches its zenith every day in the south. You learn this at primary school today, and even back then it wasn't a great mystery. People also knew that, throughout the year, the sun didn't always climb as high in the sky. It remains lower in the winter than in the summer and the amount of time that it shines each day is also dependent on the time of year. There are two days in the year, however, when the sun spends the same amount of time above the horizon as it does below it, and these are called equinoxes. These days mark the beginning of spring and autumn respectively and, from an astronomical point of view, are when the sun is exactly perpendicular to the equator. Or, to put it another way, if we were standing directly at the equator on these days, we would see the sun directly above our heads when it reached its zenith.

But only at the equator! If we are further north or south, our angle of vision shifts and the sun is no longer directly above our heads, reaching its peak slightly lower in the sky. Where this culmination (as it is called in astronomy) occurs, depends entirely on the latitude. You simply have to observe the highest point of the sun on an equinox and can then calculate how far your current position is from the equator. Should you not want to wait for an equinox, it is in principle possible on any other day, too. The calculations become a little more complicated, since you have to take into account factors like the inclination of the earth's axis, but with a certain amount of mathematical skill and a few simple observation instruments, determining your latitude is no problem. And if you want to calculate your position at night, when there is no sun to be seen, the whole thing works just as well with the stars. That was true even back then.[3]

Determining longitude, on the other hand, seemed a practically impossible task. Two things made it extremely complicated: the fact that there is no natural point of reference for longitude and the difficulty in determining the time. Latitude tells us how far north or south of the equator we are. Longitude, however, tells us how far east or west we are from the ... um ... yes, from where? That was the first problem: what does our measurement of longitude refer back to? We have no choice but to draw a more or less arbitrary line from north to south and to define this line as the point of reference for longitude. Today, we use the line that runs from the North Pole exactly through Greenwich Observatory and on to the South Pole. And it is of course no coincidence that this prime meridian runs precisely through the place where John Flamsteed worked. For if you want to measure longitude, you need to know exactly what time it is and, at least back then, only the astronomers knew this.

There were clocks of a kind in the seventeenth century, but they were sundials, of the kind that the young Isaac Newton had constructed as a child. There were also hourglasses and water clocks and even simple mechanical clockworks, which could be found in some churches. But there were no accurate clocks. You could count yourself lucky to be able to measure time to within an hour, though the somewhat more accurate pendulum clocks that had just been invented in the second half of the seventeenth century were an exception. But they were heavy and unwieldy and certainly couldn't be used on ships rolling about on the high seas.

But it is necessary to know the time in order to calculate longitude. Why this is the case can again be best explained using the sun. Let's imagine John Flamsteed, after a long night of astronomical observation, stepping outside the door of his observatory in Greenwich. The clock there shows six o'clock in the morning, and he sees the sun just rising above the horizon.

Of course, the sun doesn't really move—it just appears to do so because the earth rotates around its own axis, turning from west to east. So when Flamsteed sees the sun rising on the eastern horizon, what he really sees is how the part of the earth where Greenwich Observatory is located is in the process of rotating out of the shadow and toward the sunlight. Behind him, in the west, it is still dark. For the captain of a ship that is just crossing the Atlantic from America toward England, for instance, it is still the middle of the night. The earth still has to rotate a bit more before the captain too can see the sun on the horizon and a new day can begin for his crew.

Some six hours later in Greenwich, the sun has reached its highest point in the sky. It is now noon, and John Flamsteed will probably be unaware of this, since he will be fast asleep in bed (as active astronomers often are during the day). The captain in the North Atlantic probably can't afford to take a daytime snooze, but will also note by observing the sun that it isn't yet midday where he is, since the sun is still climbing in the sky. In his location, west of Greenwich, the earth still has to rotate a little before the sun has reached its peak there, too.

Today, we're used to there being "time zones," so that the time is the same within a certain country, or at least large areas of it. But these zones are purely arbitrary and have nothing to do with any astronomic reality. The so-called "solar noon" is not when the clock shows exactly twelve o'clock, but is instead the point in time when the sun has reached its zenith. This occurs at different times in different places, which means that the local time (or "solar time" as it is called in astronomy) is also different everywhere.

When John Flamsteed and the captain in the North Atlantic observe the position of the sun, they arrive at different results, since one of them is further west than the other and the sun is higher in the sky for one of them than for the other. In other words, the solar noon occurs at different times for each of them. And if the captain knew what time Flamsteed's clock was showing, he could use this to calculate his position. He can see how high in the sky the sun is, and also how far it is until it reaches its highest point. He knows how long this will take and he knows what time the clock at Flamsteed's location in Greenwich is showing, where the sun has already reached its highest point. And he can use this difference to calculate exactly how far west of Greenwich he is. The meridian that runs through Greenwich from the North Pole to the South Pole would be the zero point of his measurement, the so-called "prime meridian." In order to achieve his goal—finding out how far east or west of this reference line he currently is—the captain would simply have to do a few calculations.

This distance from the prime meridian is measured in degrees, with a complete circuit around the earth being 360 degrees. The captain also knows that the earth rotates once around its axis in twenty-four hours.[4] Hopefully, he will be up to the task of dividing 360 degrees by twenty-four hours and arriving at fifteen. The speed of the earth's rotation is therefore fifteen degrees per hour and so, if the sun in Greenwich has reached its highest point exactly an hour earlier than in the North Atlantic, this must mean that the ship is exactly fifteen degrees west of Greenwich.

The captain could see all of this just by taking a glance at a clock showing him what time it is in Greenwich. But this isn't possible, however, since such a clock doesn't yet exist. Back in those days, a different method was needed, and Flamsteed's aim was to provide this by creating a catalogue of the stars' positions that was as accurate as possible, thereby hoping to find in the heavens the equivalent of the clock that did not yet exist on Earth.

It would be perfect for such a celestial clock if there was an event that could be observed everywhere in the world—or at least almost everywhere—and about which one could be completely certain when it occurred. A solar eclipse, for example. One could use the eclipse to calculate on what day and at what local time it could be seen from Greenwich and then relay this information to the captain before his ship departs. When the eclipse takes place, the captain could make a note of the local time at his location, as he can himself determine by observing the position of the sun. Now he need only ascertain the difference between the local time and the previously calculated time when the solar eclipse can be seen from Greenwich and can thus calculate his position from this difference. Such a method can be used, but it isn't practical. After all, people want to be able to determine their position at all times, not only when a solar eclipse just happens to be taking place. Something is needed that is visible more often. Like the moon, for example.

NEWTON IGNORES THE CELESTIAL CLOCK AND COMMON DECENCY

And this is where John Flamsteed and Isaac Newton encounter one another again, with unhappy consequences for Flamsteed. It was his job to develop a method by which the moon could be used as part of a cosmic clock that was visible everywhere, whereby the moon is the clock's hand and the stars are the numbers showing the time. Every night, the moon traverses the sky, passing the stars as it does so. If we know the path of the moon and the positions of the stars, we can also calculate when the moon finds itself near certain stars. The captain would then take a thick book with him on his voyage, with all of these different positions listed in it. He would observe the moon and work out how far it is from a particular star. He could then look up the local time for Greenwich previously calculated for this particular constellation and compare it with his own local time.

The astronomer John Flamsteed measures the positions

of the stars and the physicist Isaac Newton calculates the trajectory of the moon using his theory of gravity. It sounds like it should have been a match made in heaven, if only Isaac Newton hadn't been such a complete egoist.

Between 1694 and 1696, the two of them worked together quite harmoniously, but Newton failed to achieve conclusive success with his work on lunar theory. After moving to the Royal Mint, he was occupied by other things and put the collaboration on ice. But when he was named president of the Royal Society in 1703, he returned to his study of the heavens.

He planned to write a new edition of the *Philosophiae Naturalis Principia Mathematica*, to include a comprehensive mathematical study of the moon's motion. For this, he needed new observation data and he intended to get these from Flamsteed. The two men met at Greenwich Observatory in April 1704. Naturally, Flamsteed was in principle equally interested in finding a practical way of implementing the lunar-distances method. After all, it was precisely for this purpose that his observatory had been built. However, he disliked Newton's approach. Flamsteed intended to release a multivolume publication that would provide a suitable framework for his decades of observations. His aim was to begin with an extensive description of the instruments and methods he had used, followed by a catalogue listing thousands of star positions, as well as his observations of the moon, the planets, various comets, and more, plus a summary of the star catalogues already in existence, stretching back to ancient times. The book, which he planned to call "Historia Coelestis Britannica," would be his life's work.

Isaac Newton couldn't care less about any of that. He was only interested in the information about the moon and the planets, and everything else was superfluous for his new edition. And if he had no need of it, then he saw no reason why anybody else should waste time on it. He had no interest in or sympathy for Flamsteed's ideas or career. He was solely interested in completing his own work and demanded that Flamsteed should assume the subordinate role, doing exactly what Newton wanted, namely providing the information that Newton needed and not making him wait for it because of some other scientific undertakings.

In order to get what he wanted, Newton started off in a cajoling tone. In a letter to Flamsteed, he wrote that the two of them would become famous if Flamsteed would provide the required observations to help him, Newton, with the implementation of the lunar-distances method. Flamsteed wasn't convinced. He wrote to an acquaintance that Newton was behaving in a "hasty, artificial, unkind, arrogant" manner, with Newton providing evidence of this arrogance in a further letter to Flamsteed: "I consider this theory to be so complex and the theory of gravitation so necessary for its understanding that I am convinced that it can never be perfected by somebody who does not understand the theory of gravitation as well as I, or better than I."[5]

Newton was fully convinced that he alone could come up with a theory for the motion of the moon and that Flamsteed was therefore obliged to provide him with his findings. But Flamsteed continued to refuse to cooperate. Before publishing his painstakingly gathered observations, he wanted first to carefully evaluate and present them. Added to which he had already had a bad experience when providing Newton with information. The previous time they had worked together, he had only given Newton his observation data on the condition that it wouldn't be passed on to others. So what did Newton do? He brazenly sent the information to a colleague, without even mentioning that Flamsteed was the originator. It's quite understandable that the latter was upset about this; today, few scientists would have any qualms about calling Newton exactly what he was: an asshole. Especially since he himself had always been aggressive and offended if he thought that somebody was using his work without suitable acknowledgment.

NEWTON PILES ON THE PRESSURE

Flamsteed certainly had no intention of letting Newton dictate when and how he should publish his findings. Newton, on the other hand, was determined not to go without the data. If Flamsteed refused to cooperate willingly, then he had to be forced to do so. Once again, Newton used the influence that

he had thanks to his niece, Catherine Barton, and her relationship with Baron Halifax, the queen's chancellor, and managed
to obtain an audience with Prince George, Queen Anne's
husband, who ordered that Flamsteed's findings should be
published as quickly as possible, even providing funds for this.

Flamsteed could hardly go against a royal edict. But he
didn't give up his plans for his life's work. In order to speed
things up, he wanted to employ two assistants to help him with
the onerous calculations that were required to turn the observation data into a useful catalogue and applied to Newton for
the necessary funding. Newton agreed to this, but demanded
that the assistants should work exclusively on the data for the
moon and the planets. The star catalogue, so important for
Flamsteed, was of no interest to Newton. Flamsteed, however,
simply told his assistants to work on the stars data, but when
Newton learned this, he immediately refused to provide the
payment.

This dispute seems childish, and in a sense, it was. If
Newton had simply let Flamsteed get on with his work, he
would have obtained the information he needed, sooner
or later. But he didn't want to wait. Unlike John Flamsteed,
Newton wasn't really interested in solving the navigation
problem and simply wanted to write an improved version of
his *Principia*. For this, he needed to use his theory of gravitation, in order to be able to describe and predict the motion
of the moon. With the help of his findings about the forces

of attraction between the heavenly bodies, this shouldn't have been a problem. Or at least that's what Newton thought at the time; today, we know that such a task was a step too far even for a genius like him. It is indeed not that difficult to work out the gravitational pull between two celestial bodies. Newton was able to do this with his theory. The moon, however, is not only influenced by the earth's gravitational pull but also by the sun's. How strong the earth's influence on the moon is depends on the distance between them, and this is constantly changing, because the earth, in its turn, is influenced by the sun's gravitational pull. Newton found no exact solution for this complex problem and we know today that he was unable to do so. Such a problem involving three bodies cannot be solved exactly in a mathematical way; one can only approximately predict the movement of the heavenly bodies, with suitably complicated methods, and Newton was not aware of these methods back then. His theory of lunar motion was therefore correspondingly inaccurate.

He did not consider himself responsible for this, however, and put the blame squarely on Flamsteed, whose observation data he claimed were too imprecise to work sensibly with. He was convinced that Flamsteed was in possession of much more precise data but was deliberately hiding this from him, and so Newton made an aggressive attempt to force the publication of this data. Flamsteed, for his part, was keeping nothing under wraps and simply wanted first to complete his observations of the sky. He wanted to chart as many stars as he could, mathematically process all of his findings, and produce a complete catalogue of the heavens—in order to solve the problem of determining longitude and to provide future astronomers with a clear and reliable basis for their own work.

But Newton didn't let up and continued to demand the data concerning the moon. A look at the correspondence between the two gives a particularly good picture of their strained relationship. Newton starts by thanking Flamsteed for some solar tables that he had received from the astronomer, but then, in the very next sentence, begins to grumble: "These and almost all your communications will be useless to me unless you can propose some practicable way or other of supplying me with Observations." Mentioning Flamsteed's weak health, he then gets straight down to what he is really concerned with: "I will therefore once more propose it to you to send me your naked observations [of the moon] and leave it to me to get her places calculated from them." At the end, Newton becomes truculent: "If you like this proposal, then pray send me first your observations for the year 1692, and I will get them calculated, and send you a copy of the calculated places. But if you like it not, then I desire you would propose some other practicable method of supplying me with observations; or else let me know plainly that I must be content to lose all the time and pains I have hitherto taken about the moon's theory."[6]

A few days later, Newton writes to Flamsteed once more to remind him that he had previously assisted him in other matters, such as the calculations for the motion of Jupiter's moons, and he takes care to point out that this "cost me above two months' hard labour." He says that he would never have taken this trouble if Flamsteed hadn't begged him to help, and also if he hadn't expected to receive help in return.

Flamsteed was determined not simply to bow to Newton's demands, however, and just gave him a few older observations and introductory texts for his planned work. He reasoned that it was okay for Newton to publish these and he himself would meanwhile have time to finish the rest of the catalogue as he saw fit. But Newton wanted all of the data. The two men came to a compromise: as a sign of trust, Flamsteed gave Newton a sealed envelope containing all of his observations so far. While the first part of the catalogue was being printed, he could continue his work and complete the data. These would then be exchanged for the incomplete data in the sealed envelope and Newton could then publish the finished catalogue.

After a long dispute, the publication of Flamsteed's data finally began under Newton's supervision. The astronomer was by no means pleased about this. He learned, for example, that Newton was paying the printer an incredible amount of money for the publication of the catalogue—more than one pound per page, while Flamsteed himself was getting no payment at all, despite only earning one hundred pounds a year as the Astronomer Royal (not enough to cover his living costs, which is why he was forced to have a second job as a country parish priest). By way of comparison, Newton, as an official of the Royal Mint, made at least 1,500 pounds per year at the time. Flamsteed complained vehemently about this situation in a letter: "'Tis very hard, 'tis extremely unjust, that all imaginable care should be taken to secure a certain profit to a bookseller, and his partners, out of my pains, and none taken to secure me the re-imbursement of my large expenses in carrying on my work above 30 years."[7]

For a time, Flamsteed's circumstances seemed to improve. When Prince George died in 1708, the astronomer considered the royal edict to be defunct and ignored Newton's demands from then on. There was little Newton could do about this to begin with, though he did at least ensure that Flamsteed was excluded from the Royal Society (for the ostensible reason that he hadn't paid his membership fee on time).

THE WORST KIND OF BOSS

Two years later, however, Newton obtained an edict from Queen Anne herself and, on top of this, managed to get himself a post as a kind of supervisor for Greenwich Observatory. Isaac Newton had thereby become John Flamsteed's boss—and he now let his mask slip fully. In letters, he more or less accused the astronomer of treason for not releasing the data: "You are therefore desired either to send the rest of your catalogue to Dr. Arbuthnot or at least to send him the observations which are wanting to complete it, that the press may proceed. And if instead thereof you propose any thing else or make any excuses or unnecessary delays it will be taken for an indirect refusal to comply with her Majesty's order. Your speedy & direct answer & compliance is expected."[8] He also opened Flamsteed's sealed envelope and started publishing the—incomplete—data contained therein, leaving Flamsteed shocked to see how his life's work was made available to the public in such a careless and fragmented way.[9]

Newton and his colleagues had messed around at will in Flamsteed's catalogue, without having consulted him. An extract from a letter that Flamsteed wrote to John Arbuthnot shows just how outraged he was. Arbuthnot was a natural scientist, like Newton, and the doctor to Prince George, and he was integral to Newton obtaining the royal edict to force the publication of the data. Arbuthnot said that Flamsteed shouldn't complain, since Newton and his colleagues had simply wished to improve the astronomer's work. In his reply, Flamsteed doesn't mince his words: "I have now spent 35 years in composeing & Work of my Catalogue . . . I have endured long and painfull distempers by my night watches & Day Labours, I have spent a large sum of Money above my appointment, out of my own Estate to complete my Catalogue and finish my Astronomical works under my hands: do not tease me with banter by telling me these alterations are made to please me when you are sensible nothing can be more displeasing nor injurious, then to be told so."[10]

Flamsteed terminated all collaboration with Newton. He announced that he would publish his data himself, at his own expense, and certainly better than Newton had done. Newton, meanwhile, refused to accept this and summoned Flamsteed to question him once more about data on the moon that he had supposedly withheld. Their conversation culminated in insults.[11] In 1712, it finally happened: the catalogue that Newton had stolen from Flamsteed was published without the latter's involvement with the title "Historia Coelestis." In his foreword, Newton dishes it out one last time, accusing the astronomer of only having worked for himself and, on top of that, of having provided poor-quality data that required lengthy corrections.

Newton was now in a position to publish the new edition of his *Principia*, including a new theory of lunar motion. Before doing so, however, he carefully erased every indication contained in the first edition of the contribution that Flamsteed had made with his observations. It didn't help him, though—his lunar theory remained inaccurate, too inaccurate at least to be used for the lunar-distances method. It was only those who came after him who succeeded: in 1753, the German astronomer Tobias Mayer and the Frenchman

Nicolas-Louis de Lacaille managed to predict the motion of the moon so accurately that it was indeed possible to use this to measure one's position at sea. By then, however, this was no longer completely necessary, since not long afterward, the clockmaker John Harrison made clocks that were compact, robust, and accurate enough to be used at sea. Nevertheless, the work carried out by Flamsteed and the other astronomers was not in vain. To begin with, the new clocks were too expensive for many sailors and it was in any case always good to have an alternative. Right up until the second half of the nineteenth century, therefore, the lunar-distances method continued to be used to navigate at sea.

For Flamsteed himself, the story had something akin to a happy ending. Queen Anne died in 1714, and Newton lost his influence at the royal court. Now Flamsteed's voice was heard and he actually managed to have the unsold copies of Newton's catalogue taken off the market. With a great sense of satisfaction, he burnt them in the observatory garden and continued with the publication of his life's work, though he died five years later before he could finish the final volume, a task reserved for his widow and two of his assistants, who completed the catalogue at their own expense. In 1725, the *Historia Coelestis Britannica* was finally published and turned out to be exactly what Flamsteed had always wanted it to be and what even Newton could not prevent him from producing: the best catalogue of stars in existence at the time and a work that would ensure that Flamsteed's name and achievements would not be forgotten even today.

So many astronomers referred to Flamsteed's catalogue in the centuries that followed that Greenwich Observatory and the meridian running through it was officially and internationally recognized in 1884 as the prime meridian for the standardized measurement of time and location. The original part of the observatory in Greenwich is now called Flamsteed House and contains a museum that pays suitable tribute to the astronomer's life and work.

THE FINE LINE BETWEEN PERSISTENCE AND HUBRIS

In his dispute with John Flamsteed, Isaac Newton revealed his most unpleasant side—vengeful, small-minded, vindictive, selfish, and so sure of himself that he sought the cause for all problems in others rather than himself. It would thus seem abundantly clear that Newton could never be a role model for a modern scientific career and could only ever set a bad example.

But things are not as simple as that. Of course, his enormous ego would cause just as many problems today as it did then—probably even more. But criticizing the work of others is a fundamental part of modern scientific methods. Research results are only taken seriously on a wider level when they have undergone a process of evaluation, in which colleagues scrutinize the methods and results involved as closely as possible. Those who aren't prepared to go along with this are not really in a position to take active involvement in the modern scientific world. Ideally, though, self-appraisal should come before criticism from others. Prior to going public with one's scientific work, one should first consider any possible ways in which one might have made mistakes. Simply blaming other people for problems, as Newton did, is by no means the answer.

Newton was unable to develop a satisfactory theory of lunar motion, but it never occurred to him that the reason for this might be found in his own methods. Instead, overestimating his own brilliance, he blamed John Flamsteed and the supposed flaws in his data. Had he focused more closely and, above all, more critically on his own work, and had he welcomed public criticism and differences of opinion with his colleagues, then he might perhaps have succeeded after all in developing the new mathematical methods that were necessary to come up with a practical theory of lunar motion.

From today's perspective, on the other hand, what is more understandable is that Newton wanted at all costs to have John Flamsteed's astronomical observations made public. If we set aside the somewhat mean motives that caused Newton's extremely ruthless approach, we must recognize that Flamsteed was the astronomer royal and not an amateur scientist (even if he did pay for practically all of his instruments out of his own pocket). Newton was therefore at least partly justified in claiming that the data weren't really Flamsteed's property and demanding their publication. In the world of research today, this is an even more pressing issue: basic research such as astronomical studies is almost exclusively funded by public money and so data gained from this research should also be publicly available. The researchers are fully aware of this, but, like Flamsteed back then, they aren't always happy about it.

A good example of this is the controversy that developed in 2014 around the Rosetta mission to the comet 67P/Churyumov-Gerasimenko. Rosetta was a project run by the European Space Agency (ESA), a body funded from the national budgets of its member states. In other words, it was taxpayers' money that allowed the mission to be carried out. The public did indeed follow the mission with interest; it was after all the first flight of its kind to a comet and the first landing on the surface of a comet, and people expected fantastic pictures and findings from the mission. Nevertheless, pictures of the comet and its surface were released only sparingly and after a long delay, with most of the photos and data being kept back or made available for use only by the scientists involved.

This situation resembled the conflict between Newton and Flamsteed. Just as Newton had demanded the immediate release of Flamsteed's observation data, space fans demanded the publication of Rosetta's pictures. And just as Flamsteed had refused Newton's request, because he first wanted to process his findings himself as he saw fit, the ESA and the research institutes involved in the Rosetta mission insisted on first evaluating the findings themselves. Both sides had justifiable arguments; the scientists involved in the Rosetta mission had, after all, spent decades working on the space probe and the various instruments. When the work finally paid off and produced new scientific results, it was natural that the scientists should first want to work with these themselves. The possibility of "fame" wasn't even a deciding factor here, as it might have been in the case of Newton and Flamsteed. Specialist publications are the currency of modern research. The more articles one publishes in scientific journals, the greater one's chances of a decent career. The scientists involved in the Rosetta mission would have considered themselves cheated out of such prospects by the premature publication of their data in much the same way as Flamsteed feared for his life's work.

The opposing argument, however, is equally valid: it is a public project, financed by public money, and basic research in particular needs the public as a lobby. The findings belong to everyone and everyone should be able to work with them. Knowledge mustn't be hidden away; the more people who have access to it, the greater the likelihood that major discoveries can be made. Newton was convinced that he was far better suited to working with Flamsteed's data than Flamsteed himself. And even if this wasn't necessarily true in that case, the general principle is not wrong. Who can say for sure that the Rosetta scientists could gain the best results from the mission's data?

At least in this particular case, a compromise was finally reached. The ESA hadn't planned to keep the data permanently locked away in any case. It merely wanted its own scientists to have priority access for a while, in line with international practice. It was nevertheless decided to publish at least some of the current pictures, so that the public could also immediately benefit from the mission's findings.

Newton's egoism and inflated view of himself shouldn't serve as a guiding principle for today's scientists. But in terms of free access to research data, his obstinacy sets a good example, to a certain extent.[12] Although Newton himself would probably have been extremely surprised to serve as an advocate for an open data initiative, as the following chapters will show.

CHAPTER 3

FALSE MODESTY AND EASILY OFFENDED: NEWTON THE SHRINKING VIOLET

$\mathfrak{T}$ruth is the offspring of silence and meditation. That's what Newton once said, and the quote makes him sound like a Far Eastern Zen master, who is at peace with himself and whom nothing can trouble. Newton certainly contemplated a lot, and he may well have discovered all sorts of truths about the universe. But calm and even-tempered—and silent? No, not at all, especially when people criticized him. This made him hypersensitive and fight back, and in the course of his career, he had countless extensive and vigorous disputes with his peers.

When Newton was still at the beginning of his scientific career, he attempted to escape criticism by remaining anonymous. The consequences of this would continue to shape his life decades later. Had he not had such a great fear of other people's criticism, many of his feuds would perhaps never have occurred.

It all began with mathematics, but the actual trigger, slightly surprisingly, was astrology. In 1664, Isaac Newton had finished his studies at Cambridge University and began to devote himself to mathematics. He had previously come across a book on astrology and had been unable to understand one of the diagrams in it.[1] There was no focus on mathematics in university courses at the time, which tended to concentrate on theology and the classics. So Newton simply taught himself mathematics, and when he started something, he only ever stopped when he had thought through every aspect of it. From self-study, therefore, he graduated onto research into topics on the very limits of existing mathematical knowledge.

IN SEARCH OF INFINITY

In Newton's notebook, we can find all sorts of calculations. Infinite series, for example, i.e., the sum of infinitely many numbers. Such sums don't always need to be infinitely great, of course; if the figures get smaller and smaller, then they will converge to a finite value.[2] In his notebook, there is a calculation in which he has worked out the value of such an infinite sum to the fifty-fifth decimal. "I am ashamed to tell you to how many figures I carried these computations, having no other business at the time," he wrote to an acquaintance in 1666. "I had too much pleasure in these discoveries." The discoveries of which he speaks were new methods of dealing with infinity. Though infinity was already a research branch of mathematics at the time, it was still considered to be rather mysterious and indecipherable. The Frenchman René Descartes, whose mathematical works Newton had studied carefully, wrote in his *Principles of Philosophy*:

> We will accordingly give ourselves no concern to reply to those who demand whether the half of an infinite line is also infinite, and whether an infinite number is even or

odd, and the like, because it is only such as imagine their minds to be infinite who seem bound to entertain questions of this sort.[3]

Well, Newton certainly seemed bound to entertain such questions. And he did so with great success. His thoughts on infinity culminated in a completely new mathematical discipline (see chapter 7), which changed the world at least as much as his findings about nature did. To begin with, however, Newton worked alone and out of the public eye. Only Isaac Barrow, who had held the mathematics chair at Cambridge since 1664, knew what Newton was up to and recognized the significance of his work. It was also Barrow who was later able to convince Newton to share at least some of his findings with the rest of the world.

In 1668, Barrow brought a book with him to the university. It was written by the German mathematician Nicholas Mercator, who had also worked on the calculation of infinite series. Newton must have been surprised and shocked in equal measure when he saw Mercator's book, for it contained precisely what he himself had thought out years before. His own work, however, was much more comprehensive than Mercator's and, at Barrow's insistence, Newton wrote up a little of what he had done.

Barrow sent the whole thing to the mathematician John Collins, who served a similar purpose in the seventeenth century to the major internet forums or mailing lists today. He was in contact with the most influential mathematicians of the day and made sure that information and new publications were swiftly and widely distributed. This didn't mean that the author Isaac Newton made an appearance, however, since Newton had only agreed to the publication of his work on the condition that Barrow would not reveal his name.

Newton avoided the public gaze. Until then, he had always carried out his research alone—discovering his new mathematical principles, investigating the nature of light and understanding gravity. But he had no idea what his scientific peers thought of all this, and he probably didn't really want to know, either. As an egomaniac, he would certainly have had no problem with public recognition, but this seems to have been outweighed by his reluctance to be exposed to criticism and the need to justify his work.

It was only when John Collins's answer not only contained no criticism, but was also full of enthusiastic praise for the work, that Newton finally agreed to make his name public. This was the first time that people outside of Cambridge University had heard of his scientific work. Collins distributed Newton's article about the calculation of infinite sums throughout Europe and began extensive correspondence with Newton—though this turned out to be increasingly frustrating.

NEWTON CONSTRUCTS A TELESCOPE

It was a curious dance that took place between Collins and Newton. Collins wanted to know as much as possible about Newton's work and to have his findings published as quickly as possible. But Newton played hard to get. He would send Collins the odd equation or the occasional solution and hint at more to come, which he would then never send. And he liked to tease a little in his letters: when Collins was interested in a table with solutions, Newton answered that this would be "pretty easy and obvious enough. But I cannot persuade myself to undertake the drudgery of making it."[4] And time and time again, he insisted, at least to begin with, that Collins should under no circumstances mention Newton's name when publishing his answers: "I see not what there is desirable in public esteem were I able to acquire and maintain it. It would perhaps increase my acquaintance, the thing which I chiefly study to decline." Newton also annotated a book by a Dutch mathematician—at the urging of Barrow and Collins—and insisted shortly before the book's publication that the sentence "extended by another author" should be added, not "extended by Isaac Newton."

Even back then, Newton could have justifiably become world-famous with his new mathematical theories. But he had no interest in public opinion or that of the scientific community. Had he not been so coy at the time, the study of natural sciences might have developed differently. He certainly would have spared himself the biggest and longest row of his career (see chapter 7).

Before things came to that, though, he started an argument with his peers for quite different reasons. At the same time as he was discovering completely new realms in the field of abstract mathematics, he also came up with a wholly practical invention that was however no less revolutionary. It was the technical counterpart to his breakthroughs in physics. Theoretical physics and astronomy today couldn't do without Newton's findings about gravity and mathematics; practical astronomy is equally dependent on the thing he constructed around the same time: the telescope.

There had been telescopes for a while, of course; they had been around for about sixty years and, in the hands of the great Galileo Galilei at the beginning of the seventeenth century, had proved that they had the potential to bring entire world systems toppling down. It was probably in 1608 that the Dutch optician Hans Lipperhey came up with the idea of combining a few special glass lenses and thus constructed the first telescope. Or the telescope might also have been invented shortly before that by Zacharias Janssen, another Dutch optician. Or maybe it was Adriaan Metius, an astronomer and mathematician. All three men claimed at one time or another to have been the first. Whichever one of them it was, the fact is that he had the idea of combining two pieces of glass. Not any old pieces of glass, of course, but rather specially ground optical lenses. The first was a so-called converging (or convex) lens. If a bundle of parallel light rays passes through a converging lens, all of the light rays converge to a single point before spreading out again. The second lens was a diverging (or concave) lens, which has the opposite effect—a bundle of divergent rays is refracted to become parallel light rays. Light that passes through the converging lens can be focused and then dispersed again through the diverging lens to form an image that can be observed. The converging lens in this case is much larger than the pupil in our eye, which only measures a few millimeters across. It can therefore gather a lot more light and so, thanks to the diverging lens, we can transform this light into an image again that our small eye can see.

A telescope is therefore nothing more than an artificial, but much larger, eye. The larger the lens, the more light it can collect and the fainter heavenly bodies we can observe. The lens telescopes of the time, however, didn't always work properly. The images were often fuzzy, and colors appeared different in the telescope than in nature. Newton knew this too, but unlike his colleagues, he had an idea why this was.

He knew all about eyes and light. After all, he had already carried out experiments on them with no regard for personal damage. His optical research was not merely limited to sticking needles into his eyes, however. Between 1664 and 1666, when Newton was laying the foundations for the greatest scientific revolution of the early modern period, he was not only busy with mathematics and gravity. He was also researching optics and the nature of light.

In 1665, he had to leave Cambridge University. The plague had broken out and the university was closed, so he was forced to return to Woolsthorpe, the small, sleepy village where he had been born. Not much happened there—there were no famous institutions, research facilities, or attractions. Today, it is of course possible to visit the house where Newton was born, but nobody was interested in that back in 1665. Not even Newton himself, though he had no choice in the matter. At least he was able to continue his research and experiments there in peace and quiet, free from external distractions.

Back in Woolsthorpe, Newton did what he had done as a small child in his attic room: he observed the light. This time, though, he did so quite differently. He carried out an experiment that is now considered to be one of the most important in the history of science—one that opened up a completely new world for him and the rest of humankind, an experiment that could hardly have been simpler and yet ended up making an entire universe visible.

Newton made a hole in the wall of a room that was otherwise fully darkened. A narrow ray of sunlight penetrated through this hole and passed through a prism, with colors then appearing on the other side. So far, so unspectacular—since glass had been in existence, there had been people who noticed that colors appeared in this glass when light passed through it in the right way. It was also known that this phenomenon could be observed particularly well when a special, triangular-shaped piece of glass—a prism—was used. But it was not known why this was the case. Or rather, the reason seemed obvious to people: they assumed the colors were produced in the prism. This was the prevailing opinion, and hardly anybody questioned it.

But Newton wasn't satisfied, particularly when he came up with a method for testing this assumption, for which he needed nothing more than a second prism. In the colorful rainbow that was created behind the first prism, he placed a dividing wall with a small hole in it that was so designed that only light of one particular color would pass through it. He thus now had just one ray of light consisting of a single color—and he then had *this* ray pass through the second prism. If the colors really did arise in the glass itself, then a second rainbow ought to have been visible behind the prism. But this wasn't the case. A ray of blue light remained a ray of blue light even behind the second prism. A ray of red light remained red.[5]

The prism didn't therefore create the colors, it separated them, Newton discovered. White light is a mixture of colors—all colors. Newton established this, too, by making the complete rainbow of rays become white light again after passing through a second prism. And he made a further discovery: the light changed direction when passing through the glass, but not uniformly—blue light deviated more than red.

It was thereby clear to Newton why the telescopes didn't work so well: each color in the white light was diverted to differing degrees in the lenses inside. This meant that there was no uniform image at the end, but rather a number of images in the different colors that were slightly adjacent to and superimposed upon each other. This "chromatic aberration" could be avoided by using mirrors instead of glass lenses. It was already known at the time that light rays could also be diverted using mirrors.[6] Up until then, however, nobody had actually managed to construct a practical and working telescope with mirrors. But then along came Newton and did just that.

When the plague disappeared after 1666 and Cambridge University commenced operations again, Newton repeated and expanded his optical experiments there. In 1668, he was then ready to build a telescope that used mirrors instead of lenses. He was extremely proud of his achievement, explaining later to the husband of his niece that he had indeed made everything himself—including the necessary tools, since "If I had waited for other people to make my tools and design things for me, I had never made anything."

Newton's telescope was a masterpiece. The reflecting telescope was not his invention, but he was the first person to understand why and how it worked in the way it did. He had investigated the scientific principles of optics and then put them into practice. His telescope was small—only about thirty centimeters long. The lens telescopes that were common at the time were much longer. They had to be, since the larger the converging lens at the front was, the further back was the point at which the light was focused. If you wanted to see more, you needed bigger lenses and therefore longer—and more unwieldy—telescopes. But even though Newton's telescope was smaller, it was still clearly better. His instrument achieved a magnification of approximately thirty-five times, almost three times as much as the best lens telescopes already in existence.[7]

Newton made the mirror himself, using a mixture of copper and tin that was later known as speculum metal. Along with the main mirror, Newton also incorporated a second, smaller mirror into his telescope. From the telescope's opening, the light first fell upon the main mirror, some thirty-three centimeters wide, being reflected there and then being diverted by the secondary mirror out through another opening, where the observer could view the image. This construction enabled Newton to achieve the maximum light yield and it was thereby possible to observe the heavens without standing in the way of the light.

RAISE THE CURTAIN FOR NEWTON'S FAVORITE ENEMY

Collins and Barrow were once again the first to hear of Newton's invention. Barrow then took the telescope to London in 1671, in order to show it to his colleagues from the Royal

Society, who were all suitably impressed. The Society's secretary, Henry Oldenburg, immediately wrote to Newton, asking him to publish a report about the new instrument. He said Newton should apply for a public patent, in order to prevent other scientists from abroad from copying the telescope and presenting it as their own work. Newton was also invited to become a member of the Royal Society.

But Newton reacted coyly once again, writing in reply: "I was surprised to see so much care taken about securing an invention to me, of which I have hitherto had so little value."[8] Had he not been asked so often, he wrote, he would have kept his telescope as a matter of private interest. But he ended up sending the appropriate information to the Royal Society for publication, again insisting on anonymity—although he did feel flattered enough by the Society's praise that he agreed to become a member. And he said that he was looking forward to sharing his "poor and solitary endeavours" in the future (which is, however, rather doubtful and can be interpreted as an expression of false modesty).

In his next letter, Newton had obviously convinced himself to finally stand by his work. He said he would not only present his telescope, but also the research that had led to its construction, it being "in my judgment the oddest if not the most considerable detection which hath hitherto beene made in the operations of Nature." So much for his modesty . . .

In February 1672, Newton sent his article to Henry Oldenburg. The title held little back: "New Theory about Light and Colours." The article was read out to the members of the Royal Society two days later, and precisely what Newton had always feared then happened: he received feedback. The members of the Royal Society set up a committee to study and evaluate Newton's work on light and colors, with Robert Hooke chosen to write the final report. Which he did—and in so doing landed himself a lifelong enemy in Isaac Newton.

Hooke was seven years Newton's senior, but, unlike the younger man, already a renowned scientist. He had been accepted into the Royal Society back in 1663 and was its first curator, responsible for all experiments. He was also assigned with the task of bringing to the weekly meetings drawings of the things he had observed with his microscope. Like the telescope, this instrument had only been invented at the beginning of the seventeenth century. And as with the telescope, there is some dispute as to who was the first person to construct it. Robert Hooke was in any case the first person to make extensive use of it in order to better understand the world of small things. His illustrations of enlarged insects, needlepoints, razor blades, or mildew[9] were so impressive that the Royal Society commissioned an entire book of them. *Micrographia* appeared in 1665 and became one of the first scientific bestsellers. While the print run only amounted to 1,200 copies, it completely sold out within just a few days.

Along with these pictures, Hooke also published a theory

of the nature of light and colors in his *Micrographia*—namely, that light arose through movement and that everything that was shining was vibrating in some way. He also said that there were only two basic colors—red and blue. When pulses of light came into contact with the eye, these two color impressions were created, and where they were superimposed on one another, "many sorts of greens" arose.[10]

This was a bit sparse for a complete theory. Hooke himself wrote in his book: "It would take a little too long to explain all of that in detail and to prove what kind of movement is responsible." And: "It would take too long to insert here how I found out the characteristics of light." This didn't prevent him from claiming that he had now explained all color phenomena in the world, however.

It's no wonder that Newton's reaction was a little unfriendly. Somebody who said he knew everything, but wouldn't say what exactly he knew or how he had attained this knowledge, was only likely to irritate somebody like Newton, who had after all stuck needles into his eyes in order to understand how light worked. Particularly since Newton had himself carried out experiments during his exile in the countryside which in his opinion clearly demonstrated that Hooke's theory of colors could not be correct.

Hooke, for his part, claimed that he himself had already carried out the experiments that Newton described in his article, and this much earlier, and that Newton had also falsely interpreted the findings from the experiments. Newton was outraged at this. For him, the "most considerable detection" announced in his letter was not merely a simple observation. It was a mathematical certainty that he was presenting here to his colleagues. It was not a hypothesis, not an assumption, but rather the beginning of a new, mathematical understanding of optics. He had found the ultimate solution to one of the mysteries of nature and this "without any suspicion of doubt" as he wrote to Secretary Oldenburg. He had drawn up a hypothesis and developed an experiment that confirmed his statement about the nature of light and colors. Hooke's theory, on the other hand, was "not only insufficient, but in some respects unintelligible."

Newton seemed incapable of putting himself into the position of his peers. Yes, his explanation of light and colors was indeed (largely) correct. But he had come up with it years before and had had more than enough time to consider it and improve it. In the same period of time, he had developed his new theories of mathematics and physics and was accustomed to having completely new and innovative ideas which went far beyond what was known to the rest of the world. Now he was firing these revolutionary ideas without warning at the rest of the world, and he shouldn't have been surprised that it would take a little time before people could appreciate them. He should have been expecting criticism at first, but he wasn't. His first appearance in the scientific public eye must have seemed to him to be quite a disaster.

JUST LIKE IN KINDERGARTEN

Over the next few years, there was lively correspondence between Newton and the members of the Society. Ten critical articles were published in the *Philosophical Transactions* and eleven times Newton sent angry replies to this criticism. He was particularly upset that his discoveries about light were always referred to as "hypotheses." And he was extremely irritated when Hooke was the author. For Newton, his optical theories were mathematics and not some goddamned hypothesis. He refused to accept such a claim from anybody—particularly not somebody like Hooke, who had absolutely no understanding of the true nature of light. Newton was not prepared to permit any criticism of his work, even when it might have been valid for once. The Dutch scholar Christiaan Huygens, for instance, noted that white light is not only obtained when all the colors are mixed; two colors are sufficient, as long as they are the right ones, like blue and yellow.[11]

Huygens was right, but that didn't diminish Newton's rage against the Dutchman. He took criticism as badly as a small child, and behaved just as sulkily. He even announced that he would leave the Royal Society, in order to "avoid such things in future," as he wrote.

He didn't actually make good his threat, of course. Instead, he wrote another article about optics, containing his thoughts on the true nature of light. As defiant as ever, he specifically titled it a "hypothesis," one "explaining the Properties of Light, discoursed of in my several Papers." But regardless whether he was producing mathematical certainties or hypotheses, criticism still only provoked antagonism in Newton. His dispute with Robert Hooke entered its next phase: Hooke believed that light consisted of waves, while Newton thought it was a current of particles. Hooke said that he had done an experiment that showed the wave nature of light. Newton explained that this experiment had been carried out by other scientists much earlier and Hooke shouldn't call it "his" experiment. Hooke accused Newton of having copied out the hypotheses in *Micrographia* without stating his source. In other words, the two of them behaved as though they were in kindergarten.

The question as to whether light is a wave or consists of particles was definitively answered, not by either Newton or Hooke, but instead a few hundred years later by a new revolution in science—the development of quantum mechanics. Meanwhile, Newton and Hooke did pause in their quarrelling—but not for long, with the next round being even fiercer. Newton also stopped publishing his works for the time being, and his mathematical writings, which had been intended for publication after the works on optics, remained unknown to most of his peers, while he kept his work on gravity completely to himself. Newton's first exposure to the criticism of his colleagues seems to have discouraged him so much that he preferred to say nothing more, rather than face public judgment.

His work was indeed ingenious and revolutionary. But hardly anybody learned anything about it because he had no desire to be criticized—in other words, because he was a shrinking violet.

GOOD CRITICISM IS HALF THE BATTLE

In the world of science today, Isaac Newton's attitude would be disastrous. Constant review is the foundation on which modern research is based. Naturally, it isn't always a pleasant experience to expose oneself to criticism, but it is nevertheless indispensable. What the Royal Society did with Newton's article about his theory of light and color is now a standard process and is called a peer review.

The actual research work still forms the basis of all scientific findings. But in order for these to be taken seriously, they need to be published. And before this can happen, they need to be exposed to a review. If you want to publish scientific results, you have to submit them to a specialist publication, where an editor will first decide whether the research in question is sound or just complete nonsense. This initial review is based on rather formal criteria: does the subject fit in with the publication's focus? Is the text sufficiently well-written? Have the authors kept to the standard structure, presented their methods and findings in a suitable way, and provided appropriate sources for their assertions? And so on and so forth. Once an article has passed this first test, the actual peer review follows.

The editor chooses one or more reviewers. These are just normal scientists who work in the same field as the authors of the article in question. They are experts in the type of research that is being presented (or at least they are supposed to be) and it is their job to test the detail of the article's content. The reviewers then take a very close look at the work and attempt to reproduce the results. They check for mistakes and indicate when, in their opinion, there are unresolved questions, flaws in the methods, or problems with the conclusions. According to what they find, they then advise the editor to accept or reject the work for publication. The authors typically receive a detailed list of questions and suggestions for improvement, and these are to be answered and implemented. If they can do this to the satisfaction of the reviewers and the editor, the work can then be published.

This peer review process is designed to minimize the number of false research results that are published. Of course, it doesn't always work to perfection. Even the best reviewers can overlook mistakes. Or they don't take enough time for their assessment.[12] The subject is often so specialized that there is barely anybody who knows enough about it, and those that do are generally either colleagues or direct competitors of the authors and thus not fully objective. Despite all these flaws, however, the peer review process is still indispensable in modern science. If you don't expose yourself to this process, you can't have your work published. And those who don't publish any work don't really count—"publish or perish," as the saying goes.

Scientific careers are judged in terms of your publication lists. The more you have published, the better your chances of getting a good job or more funding. It's not an ideal state of affairs, of course—there are plenty of other criteria that ought to be taken into account: dedication to teaching, for example, or commitment to public relations work. But as long as having work published is the ultimate benchmark, then it is necessary not only to be able to tolerate your peers' criticism—you need to actively seek it.

Besides publication in specialist journals, attending symposia is also part and parcel of a natural scientist's life. Here, you can present your work to your peers, mostly with a lecture, which is inevitably followed by the part where the audience can ask questions or pass comment. The peer review process is mostly impersonal and anonymous; the authors don't generally know the names of their reviewers. The Q&A session following a lecture, on the other hand, is extremely personal and public. Many young scientists fear their first appearance at a major conference—not always without justification. Rivalries and bitter animosity were not only a feature of Isaac Newton's time—today as well you can witness fierce quarrels at conferences. Some people are kind enough to give you their feed-back in private, but others have no qualms about provoking an argument in front of the assembled crowd.

The criticism isn't always justified, but in most cases one would do well to at least think about it. The real aim of the peer review isn't to discover actual fraud or deliberately manipulated results. It is to check for the mistakes that all people inevitably make. Ideally, science should be completely objective, but we do find it incredibly difficult to think in this way. We are subjective creatures—that's just the way it is—and perhaps we just overlook certain mistakes because the results seem to fit in so well with our ideas.

Years ago, for example, I was investigating the planet system of the Beta Pictoris star. That is to say, I was investi-gating the question whether there can actually be any planets there. I was keen to deduce the existence and characteristics of possible planets based on certain conspicuous features in the observation data available. I then carried out comprehensive computer simulations and was delighted to finally be able to explain all of these features with the presence of a single planet. It was an elegant solution to the problem and I was looking forward to writing up my findings at last and publishing them in an article. First, however, I sent the whole thing to a col-league and asked him for his opinion. He replied a few days later by email: "Something isn't right—the results look too good to be true. I'd be surprised if they really are correct." My initial reaction was similar to Isaac Newton's—I was annoyed and angry that I had even bothered to show him my work. What did "too good to be true" mean? Of course the results were okay—if somebody was wrong, it was him and not me! Nevertheless, I had another look at the work and was forced to admit that I had indeed overlooked a small and insignificant flaw in my computer program. A flaw that would normally have been of no consequence but had here produced the exact result that I had wanted. It is precisely for such cases that the peer review process is required. Scientific procedures are designed to protect us from ourselves, as far as that is possible. But this is only made possible by constant criticism, which we have to seek out, listen to, and finally also accept.[13]

Sometimes, though, it really is difficult to take criticism. Even the best scientists can have problems doing so. Albert Einstein was actually an affable man. But there was at least one occasion when he was just as defiant as Newton when it came to criticism of his work. In 1936, he and Nathan Rosen wanted to publish an article about the existence of gravitational waves. In a publication twenty years previously, he had predicted that accelerated mass gives off energy in the form of gravitational waves. In his new article, however, he now came to the conclusion that gravitational waves didn't exist after all. He submitted the article to the *Physical Review* journal, and the editor sent the text to a reviewer. The reviewer concluded that Einstein and Rosen had made a mathematical mistake.

Einstein then wrote to the editor: "We (Mr. Rosen and I) had sent you our manuscript for publication and had not authorized you to show it to specialists before it is printed. I see no reason to address the—in any case erroneous—comments of your anonymous expert. On the basis of this incident I prefer to publish the paper elsewhere."[14]

He was so upset by the reviewer's comments that he decided never again to publish work in the *Physical Review*, a vow he kept until the end of his life. The article on gravitational waves was published one year later in another journal. However, it appeared in a corrected form and with the conclusion that gravitational waves could actually exist. Einstein had obviously taken the criticism to heart after all and let go of his uncharacteristic defiance.[15]

Those who wish to work in science today have to be able to live with constant evaluation by their peers. If you can't take criticism, then you're in the wrong profession. Those who want to be successful have to both seek and accept other people's judgment. Being a genius like Isaac Newton is highly useful, but it doesn't shield you from criticism—on the contrary—and Newton himself was forced to realize this when he next entered into dispute with Robert Hooke.

CHAPTER 4 GRAVITY WITHOUT THE APPLE: NEWTON'S PUGNACIOUS SIDE

Isaac Newton came up with the theory of gravity when an apple fell on his head. This occurrence is one of the great stories in the history of science, along with Galileo Galilei throwing things from the Leaning Tower of Pisa, or Archimedes running naked through the streets, shouting "Eureka!" after he had discovered the principle named after him in the bathtub.[1] Today, we know that Galileo didn't carry out his experiments on a tower and that Archimedes probably didn't carry out research in a bathtub. And it was different with the apple, too. Above all, though, we do know that the development of Newton's theory of gravity was accompanied by no end of quarreling with his archenemy Robert Hooke.

UNCERTAIN KNOWLEDGE ABOUT THE UNIVERSE

With gravity in Newton's day, it was a bit like it was with light. Nobody really knew what it actually meant. Of course, people realized that things fell downward and that there must be a reason for this. But what was it that made things fall? What made the moon always move around the earth, and the earth around the sun? And what actually was motion?

It is difficult to imagine how scientists of that time must have felt. In view of the enormous number of unanswered questions, it was difficult to establish anything with any certainty. Along with all the knowledge about the universe that we take for granted today, an entire language was missing. If somebody uses the word "gravity" today, it's pretty clear what they mean by it. Perhaps not everybody understands the specific scientific significance of the term, but most people know at least that we're talking about a force that acts between all objects with mass.

The word "gravity" was already in use in Newton's time, but it had a different meaning from our understanding of it today. Like "light," nobody at the time really had any idea what "gravity" actually was. In his notebook, the young Newton imagined it to be something material—something that was contained in all matter and caused a certain "massiness." A kind of "gravitational liquid" that was contained in everything and made things heavy. Some people thought of gravity as a sort of radiation; other contemporaries of Newton's believed it to be a place, some special location in the universe to which all objects were attracted. They thought there was a tendency within matter to move toward this location, and since everything always fell down to the ground, this special place could only be located at the center of the earth, which thus had to mark the center of the universe. This reasoning led to problems, however—what about all the other heavenly bodies? Wouldn't a stone that was dropped on the surface of the moon also move in the direction of the earth? Surely there couldn't be two such special locations in the universe? And if gravity really was a kind of liquid, what happened to it? Logically, it could only ever move downward and so must gather, as Newton wrote in his notebook, in giant hollow spaces in the earth's interior. So you can see that there were at the time all sorts of conflicting ideas and concepts.

The question of "gravity" is closely connected with the question of the general nature of motion. Things fall downward—that much was clear to Newton. But not always—a cannonball first flies upward in a wide arc, before it then falls toward the earth's surface again. The water in the oceans rises and falls with the regular cycle of the tides. There had to be an explanation for all of these different motions and a language with which to describe them.

During his studies, Newton regularly read about the experiments and theories of his predecessors, for example Galileo Galilei's reports about the speed at which objects fall to the ground, about which Newton noted down the following: "According to Galileus [Galileo] a iron ball or 100 Florentine (that is 78 lb. in London of Adverdupois weight) descends an 100 braces Florentine or cubits (or 49.01 Ells, perhaps 56 yds.) in 5 seconds of an [hour]."[2] Florentine pounds, London avoirdupois, Ells? The only unit in this list that we still commonly use today is the second.[3] This entry from Newton's notebook is further evidence of how different from today the situation was for scientists back then. There were no clear and generally accepted definitions for words like "motion," "space," or "gravity." There were no uniform measurements and generally not even practical measuring instruments. But at least there was Isaac Newton, his burning desire to understand everything, and the famous apple.

AND SO IT FELL . . .

If we are to believe William Stukeley, a friend of Newton's and the author of a biography that was published back in 1752, then the famous apple really did exist. It didn't fall on Newton's head, however. Instead, with no sense of a good story, it simply fell to the ground, as apples sometimes do. Unlike all the other ones that throughout the centuries had fallen from apple trees all over the world, however, in the case of this particular apple there was a genius sitting nearby who happened to have nothing better to do than to ponder the nature of the world, since the university where he should have been working had been shut down because of an outbreak of the plague. When Newton was taking involuntary leave in his home village of Woolsthorpe between 1665 and 1666, he didn't only busy himself with needles and the nature of light (see chapter 1), but also with mathematics (see chapters 3 and 7). Practically all of his other revolutionary ideas were conceived during that time, and none of them were as revolutionary as those about gravity.

So there was Newton, sitting under, next to, or at least within sight of an apple tree, when he saw an apple fall (or at least later claimed to have seen it). Whatever may have actually happened—Newton pondered deeply about falling and moving objects. Not just humdrum apples, but also celestial bodies like the moon. Apples move toward the earth, he thought to himself. And the moon moves around the earth. What if there was a single cause that was responsible for both phenomena? What if the earth exerted an influence that affected both the apple and the moon? What kind of influence could that be? The moon is much larger than an apple. But the apple is much closer to the earth than the moon. The moon "hangs" above the earth; the apple hangs on the apple tree. The moon moves around the earth, while the apple hangs or falls down in a straight line. There seemed to be some sort of connection between all these thoughts, but what exactly was it?

Viewed from a distance of a few meters, an apple can appear to us to be just as large as the moon. That's logical, since the closer something is to our eyes, the larger the perceived image. Whatever is true for the apple must also be valid for the moon. If the moon were therefore twice as close, would it then appear twice as large? No—Newton knew that the surface area of the moon would then appear four times as big. And if the moon were four times closer, it would look sixteen times bigger. The appearance of the surface area does not therefore appear proportionally greater as the distance gets smaller—it changes exponentially. Mathematically, this is referred to as the inverse-square law.

The tiny apple and the gigantic moon seem to be equally large, therefore, but only because the apple is much closer and the appearance changes not in proportion with the distance, but is rather inversely proportional to the square of the distance. Perhaps this is also true of the influence that Newton is trying to find? Using the limited data available to him at the time about the distance and size of the moon, he made a small, rough calculation. And it was indeed the case that an influence that was inversely proportional to the square of the distance was suitable to affect both the apple and the moon.

Today, we like to imagine the conception of the theory of gravity as a typical flash of inspiration, a spontaneous and instantaneous brainwave that was set off by the apple falling. In reality, it was a longer and much more complicated process. The ideas that Newton had in the Woolsthorpe orchard merely formed the basis for his later scientific work. There was still a long way to go first—and plenty of trouble, too.

First of all, Newton returned to Cambridge. He studied mathematics, constructed his telescope, published his first findings about the nature of light and color, and got involved in his first dispute with the members of the Royal Society. After all of that, he devoted himself to completely different subjects (see chapter 6). He only returned to gravity in the 1680s, and the catalyst for this was a comet.

THE WRATH OF THE COMET

In the seventeenth century, as might be expected, comets were still largely a mystery. Nobody knew what they consisted of. Nobody knew why they shone so brightly. Nobody knew why they always appeared so suddenly in the sky and then disappeared again just as suddenly. What people thought they knew at the time was nonsense.

When a comet appeared in the sky at the end of 1680, there was great excitement. It was described as a "sword of vengeance and rod of wrath of Almighty God" in a pamphlet in Nuremberg, which spoke of "a terrifying torch, rod, and sword as a benevolent warning of pending doom."[4] The comet was seen as a warning from God, designed to bring about "dread and transformation in hardened sinful souls." But not only in Nuremberg were people convinced that comets were the harbingers of doom. This view had been held all over the world for some time. Unlike the stars and planets, the motion of the comets was irregular and unpredictable, and their appearance was out of the ordinary. They weren't points of light that shone brightly and steadily, but rather dull clouds without regular form and with tails that could stretch across the whole sky. The Roman historian Pliny the Elder had already explained that the appearance of a comet signified a looming disaster in his *Natural History* from the year 77 CE, and up until Newton's day, this view had scarcely changed.

Hundreds of scripts, leaflets, religious pamphlets, and other texts were written about the comet of 1680 and all of them outdid each other in painting the most terrible picture of the future.

Isaac Newton also observed the comet almost every night, eagerly following its traversal of the sky, as did the Royal Society's other natural scientists: Edmond Halley (then aged just twenty-one),[5] Robert Hooke, and of course the Astronomer Royal John Flamsteed. When the comet first appeared in the sky in November 1680, though, Flamsteed missed it. It was only visible shortly before dawn at that time and soon disappeared completely from the sky. Flamsteed conjectured that comets didn't perhaps behave as unpredictably as was thought, and forecast that it would soon return. And so it was: in mid-December, the comet could again be seen shining brightly in the night, although not everybody was convinced that it really was the same celestial body as the one a few weeks before.

Flamsteed wrote a letter to a friend at the University of Cambridge, in the hope that he would pass it on to Isaac Newton. The two men's major dispute (see chapter 2) was still a thing of the future and Flamsteed was interested in Newton's mathematical expertise. He had developed a (relatively vague) theory about the motion of comets and wanted to hear Newton's opinion. What if, Flamsteed thought, the sun attracted the comet in some way, perhaps magnetic? The comet would move in a straight line toward the sun and, when close enough

to it, would be repelled again, just as magnetism can sometimes attract and sometimes repel.

Newton wrote back and explained outright to Flamsteed
that his hypothesis was nonsense. He said the sun was hot and
it was a well-known fact that objects lost their magnetic properties when they were heated up. In addition, Newton himself
wasn't quite certain whether comets were not perhaps actually two different objects. His own records showed that the
motion of the celestial body (or bodies) wasn't consistent,
being sometimes faster and sometimes slower. But, Newton
wrote to Flamsteed, even if the comet didn't move in a straight
line toward the sun and then away from it again, it would be
possible that it followed a path *around the sun*. That could
also be caused by an exclusively "attracting force in the sun."
"Force" was another word whose scientific meaning had not
been clearly defined at the time. But Newton's work on the
motion of the celestial bodies was to change that once and for
all. He pondered deeply about forces emanating from the sun,
the earth, the moon, and other bodies in the cosmos. He wondered how these forces might be mathematically described
and in what way they were interdependent. He came to the
conclusion that the motion of the comet could be a simple
juxtaposition of two forces: first, its tendency to move along a
straight line, and then the deviation from this straight line by
a force of attraction emanating from the sun.

HOOKE V. NEWTON: ROUND 2

The comet had inspired Newton's first concrete, and above all public, statement about the manner and cause of celestial bodies' motion. It was also the catalyst for another altercation with Robert Hooke. For Newton wasn't the only one to ponder the nature of the motion of planets and comets. Hooke had similar ideas and had told Newton about them back in 1679. Wanting to settle their dispute about optics, he asked Newton for his opinion on his theory. Hooke had also had the idea of explaining the motion of planets by a force of attraction, the strength of which changed with distance. He didn't know that Newton had had exactly the same idea back in 1666, long before Hooke (he couldn't have known, since Newton had as usual told nobody about his thinking). In his response, Newton acted as though he was unaware of such a theory. But he did get involved in an exchange regarding a thought experiment.

The question sounds very simple: What happens when you drop a ball from a high tower? Naturally, it falls downward—this much was obvious and, even for an argumentative pair like Isaac Newton and Robert Hooke, there was no reason to fall out over that. But where does the ball land? The earth rotates on its axis, once a day, from the west toward the east. If, therefore, the earth under the ball rotates eastward, then the ball should land a little to the west of the tower. At least one might think so—but Newton had other ideas. The top of the tower rotates with the earth, and the further it is from the ground, the more quickly it rotates.[6] That means the ball has a greater starting velocity toward the east than objects on the ground. It must therefore land to the east of the tower, Newton believed. And not only that: in a diagram, he illustrated what would happen with the ball if it could fall unobstructed toward the center of the earth. If the motion of the ball is constantly only downward, toward the center of the earth, then the rotation of the planet ensures that it approaches the center along a spiral-formed path.

Nonsense, said Hooke—and he was completely right. His own suggestion was that the ball falling through the earth would behave like a planet that moves around the middle of the earth, probably moving, he thought, along an elliptical path. Accusing Newton of having made a mistake was enough to get a rise out of him. The fact that such an accusation came from Robert Hooke only exacerbated the problem. But what probably made Newton angriest was the fact that Hooke criticized his idea before the members of the Royal Society. The two of them had actually agreed to keep their exchange private, but Hooke broke this promise when he read out Newton's letters in public.

The old quarrel between the two scientists was now in full swing once more. Newton sketched a new diagram that showed clearly what the ball's elliptical path would look like, since Hooke's suggestion was also not completely correct and, above all, mathematically rather vague. The two men continued to argue for a while what the connection between the path of one body and the force of attraction of another might look like. Hooke's final conjecture was that the strength of the force of attraction was inversely proportional to the square of the distance, and he asked Newton for an answer to the following question: What path would be followed by a celestial body on which such a force was exerted? Hooke was aware of Newton's mathematical skill and so threw down the gauntlet with this problem. But Newton preferred to remain silent.

It took a few more years for the next important step in Isaac Newton's scientific revolution to be taken. In 1684, Edmond Halley, Robert Hooke, and Christopher Wren[7] met in a coffee house and discussed—once again—the question of the motion of objects. All three of them believed that the force that makes the celestial bodies move around the sun ought to be inversely proportional to the square of the distance. But this was just a supposition; they lacked a rigorous mathematical interpretation. Over coffee, Robert Hooke claimed, however, that he had long since demonstrated how the motion of all celestial bodies could be explained by a force that was inversely proportional to the square of the distance. He simply wished to keep his findings secret and to only go public with them when they would be truly appreciated.

Halley wasn't convinced that Hooke wasn't simply bluffing and preferred to ask Newton once more. He visited him in Cambridge in August 1684 and once again posed the question:

What path is followed by a celestial body on which a force of attraction is exerted that is inversely proportional to the square of the distance? Newton's response was immediate: an elliptical path. He said that he had long since established this using mathematical calculations. Unfortunately, he had lost these calculations, but he would repeat them and send them to Halley.

THE SILENCE IS BROKEN

Astonishingly, that is exactly what Newton did. Unlike previously, he actually delivered what he had promised. First, Halley received a short manuscript titled "On the motion of bodies in an orbit." But that was merely the beginning. Newton kept on writing, devoting himself to comets, planets, and the moon. He gathered data, asking the astronomer Flamsteed (their major dispute was still to come) for observations that he could use for his calculations. He developed a new language, providing definitions for words like "space" or "time." He used the new mathematics he had developed years before (of which hardly anybody apart from himself yet knew) to explain not only the motion of the planets, but also numerous other phenomena in the skies and on the earth. Just like that, he created an entirely new approach to natural sciences, or a "natural philosophy," as it was still called at the time. Above all, he developed mathematical principles that formed the basis of this new natural philosophy. These were the "Mathematical Principles of Natural Philosophy," which was the title he gave to his work: *Philosophiae Naturalis Principia Mathematica*.

The initial brief treatise was enough to convince the members of the Royal Society that something truly new and significant was to be expected here. They agreed to print and publish Newton's book, when it was finished, and commissioned Edmond Halley to supervise the matter. It is thanks to him that the book did indeed appear in 1686, since the project was almost scuttled because of a few fish. Or to be more precise, because of a work by the naturalist Francis Willughby, whose name has slipped into obscurity today. He wrote books about birds, insects, and fish (and football too, rather surprisingly).[8] He had died in 1672, and his works were only published after his death. In 1686, the Royal Society decided to publish Willughby's *History of Fish*.[9] The book contained a great number of illustrations of fish, which made the printing process extremely expensive—without, however, boosting sales. This history of fish was very much a shelf warmer and the Royal Society was reluctant to spend money on printing another book,[10] which meant that the publication of Newton's work was on the verge of being abandoned because of a fish lover. Without further ado, however, Halley defrayed most of the printing costs out of his own pocket; posterity and the entire world of science have reason to be grateful to him. His contemporaries repaid him in a rather curious manner, on the other hand. An agreed fee of fifty pounds for various tasks carried out for the Royal Society was withheld. Instead, he was offered fifty copies of the *History of Fish*.[11]

A SMALL THEORY FOR A BIG UNIVERSE

Besides all of these problems, Halley also had to take care of the dispute between Hooke and Newton, which had broken out once more shortly before the final publication of the work. At the request of the Royal Society, Halley wrote a letter to Newton in April 1686 to thank him for his work on the *Philosophiae Naturalis Principia Mathematica*. In it, however, he also mentioned that Robert Hooke had a slight problem with the manuscript for the first volume of the work. Hooke was of the opinion that the idea that a force was exerted between the celestial bodies, with its strength inversely proportional to the square of the distance, originated with him and Newton should mention this in his book.

It should now be clear to the reader how Newton would react to such accusations. In an enraged reply, he pulled no punches: "Now is not this very fine? Mathematicians that find out, settle & do all the business must content themselves with being nothing but dry calculators & drudges & another that does nothing but pretend & grasp at all things must carry away all the invention." In a letter to Halley, Newton attacked Hooke directly: "Mr. Hook has erred in the invention he pretends to & his error is the cause of all the stir he makes." Newton was prepared to admit that Hooke had told him about his thoughts on gravity. But he, Newton, had not asked about them and therefore felt no obligations in connection with them. Hooke had imposed the exchange of ideas upon him and Hooke's errors in his description of gravity had been the catalyst for Newton's own work and discoveries concerning the subject. Newton added that it was absurd for Hooke now to claim that all of this had originally been his own work. He called Hooke "a man of a strange unsociable temper" and promptly set about deleting every mention of Hooke from his manuscript.

As he so often did, Newton reacted far too aggressively and far too quickly, though in this particular case, he wasn't completely wrong. Of course there were other scientists at the time who were thinking along the lines of a force of attraction that was inversely proportional to the square of the distance. But these were merely ideas and suppositions. Nobody had developed a mathematical foundation or a unified and large-scale view of the world from which the law of gravity must necessarily follow. Nobody, that is, apart from Isaac Newton.

For that is what is truly revolutionary about his work. Not the discovery that one can explain the motion of the celestial bodies when one presupposes a force of attraction that is inversely proportional to the square of the distance. That was by no means the only realization contained within the three-volume *Philosophiae Naturalis Principia Mathematica* (see chapter 5). Newton created a completely new view of the world—not for nothing is the third and final volume of his work called *De mundi systemate* (Of the system of the world).

Newton had realized that the force that is responsible for the motion of the celestial bodies is not limited to the sky, but is rather a fundamental force, one that is exerted throughout the entire cosmos—on apple trees and entire stars alike. It is difficult to imagine the creative achievement that was necessary for Newton to come up with the idea of such a universal force of gravity—and not to leave it there, but rather to then explain it in mathematical terms.

It's not a question of Newton having "invented" or "discovered" gravity. He showed that it was possible to understand gravity, and that it is a universal force. He created a formula—a single theory that can be used to explain an entire universe.[12] What Newton did at that time formed the basis for modern science. He showed that the world can be understood and explained, that there are *natural laws* that apply everywhere. He took the first step toward the unification of the natural sciences and started a process that is still going on today. When modern quantum physicists search for a theory of everything, or when they attempt to unify quantum mechanics with the theory of relativity, they are able to do so because Isaac Newton started it all off all those years ago.

Newton clearly demonstrated that gravity is a universal

force, that it acts as much in our daily lives as it does on the great cosmic stage of the stars and planets. He showed the natural scientists that the world need not only be understood piece by piece and that there are links between the different, seemingly isolated phenomena around us, and that it is possible to conceive of these links in mathematical formulae and understand them.

GO IN SEARCH OF DISPUTE!

But was all that quarrelling really necessary? Couldn't Newton have settled his differences with his colleagues in a slightly more peaceful way? One thing is clear: it's impossible to avoid disputes in science. That's because scientists are completely normal human beings with human characteristics. And even if your typical scientist doesn't have as many unpleasant character traits all at once as Newton did, he or she generally has enough for the odd confrontation. And confrontation is also absolutely indispensable for science to function properly. That's down to the very nature of the matter—the role of science is to make two diametrically opposed concepts compatible.

On the one hand, of course, we always want to find out new things about the world. That is after all the very reason why we get involved in science. There are heaps of things that we don't know, and that's what we want to learn more about. And because our knowledge is incomplete, the same is true for the models that we use to explain the world. Each new model must therefore correct certain assumptions of the old ones—and in some cases, it can completely push them out of the picture.

On the other hand, however, there is also the tendency during research to stick to what you already know. In order for something to become a part of accepted science, it has to actually *work*. That's true of Newton's model of gravity, for example. His equations enable us to predict the movements of the celestial bodies. We can use them to steer space probes so accurately through the solar system that they can land precisely on planets that, when the probes were launched, were still in a completely different location. Newton's theory of gravity works—even though we know today that it is "wrong." In 1915, Albert Einstein developed a completely new model of gravity. With Einstein's equations, we can calculate everything that we can with Newton's mathematics, but we can do so more accurately and also in cases where it wasn't possible before. One could say that Einstein works better than Newton—which doesn't change the fact that Newton still works. At least, he does within certain limits, and thanks to Albert Einstein, we now know exactly what those limits are. The greater accuracy in the general theory of relativity comes at the cost of greater complexity; it is much more onerous to calculate using the new equations than the old ones. And because the old ones are still good enough within their limits, we still use them as well today.

A certain caution, however, is generally a good thing when

it comes to new scientific theories. A great deal of the research being undertaken on the outer limits of current knowledge is by definition doubtful. One speculates and makes assumptions, and doesn't always have the chance to immediately test such speculation with concrete experiments. If scientists were to reject the status quo straightaway every time anybody had a new idea, then nothing more would be achieved. An accepted scientific theory has been checked in countless experiments and tested using countless observations. Such hard-won theories shouldn't then be rejected just like that, and this somewhat conservative attitude is essential. At the same time, however, science must be open to new concepts and prepared to entertain ideas that nobody has ever had before.

It is logical that conflicts should arise, therefore. If you have a new idea, you are generally fully convinced by it and have no time for existing knowledge. That's where your peers come in— their job is to view and judge the potential advance as soberly as possible, in order to see whether it is really a revolutionary idea or simply wishful thinking. Such a situation can easily give rise to disputes—and these disputes are sometimes necessary.

Science isn't democratic. What happens isn't necessarily what the majority wants, and the truth is not always to be found in the middle. Even if all of your colleagues and peers are ranked against you, it is still possible that you are right. In which case, you mustn't allow the waves of criticism to overwhelm you—and you have to be prepared for animated discussions and disputes. On the other hand, we mustn't take this idea too far. Just because you face criticism, it doesn't automatically follow that you are right.[13] There are reasons for sensible criticism, and you have to take these reasons seriously. And it isn't necessary to let every animated discussion escalate into an argument. But if you want to be a successful scientist, a certain amount of pugnacity isn't necessarily a bad thing. You need to be able to stand up for your own ideas and not to throw in the towel at the first sign of criticism. Equally, you need to be able to criticize other people's ideas, where appropriate, and not to accept everything simply because somebody claims it is true. Science requires from all involved the willingness to engage in vigorous discussion; otherwise it cannot function properly. You just don't need to take this to extremes as Newton did.

Regardless of any readiness for an argument, in one respect, you definitely shouldn't take Isaac Newton as an example. You should recognize the established achievements of other people, even if you cannot stand those people. Simply to ignore their contributions—as Newton did with Hooke (and also Flamsteed or Leibniz—see chapters 2 and 7)—is a serious mistake. It has nothing to do with confrontation and criticism any more. It is at best unnecessarily impolite and at worst a clear case of plagiarism, which means presenting the thoughts and findings of others as your own. Today, this is quite rightly considered to be a serious malpractice—even if you are as brilliant as Isaac Newton.

CHAPTER 5 THE SILENT REVOLUTIONARY: NEWTON THE MYSTERY-MONGER

$\mathcal{J}$f Isaac Newton had achieved nothing else in his life other than finding a mathematical demonstration of gravity, that would still have been enough to ensure he went down in history as one of the most important scientists of his time. The *Principia*—finally published in 1687 in spite of all the difficulties and a dreadfully expensive historical treatment of fish—has three volumes, however, and these contain much more than just a few ideas about gravity.

ON THE SHOULDERS OF . . .

The physics Nobel Prize winner Steven Weinberg said the following about Newton's work: "All that has happened since 1687 is a gloss on the *Principia*,"[1] which may be a bit over-the-top, but only a bit. Despite its fundamental and timeless significance, however, the work is as difficult to read today as it was back then. The fact that the *Principia* is such an onerous read is not only due to its (mathematical) contents, but is also thanks to Newton himself, since he deliberately made sure that his work was incomprehensible.

Nevertheless, it is worth taking a closer look at the *Principia*, which goes beyond the law of gravity and, for instance, deals with the following question: gravity is a force of attraction, but if it is indeed the case that every object in the universe attracts every other object, why doesn't everything end up in a giant heap somewhere in the cosmos? Why are there planets, moons, and comets moving along their paths around the sun in a somewhat orderly fashion? In order to answer these questions, Newton didn't only need a mathematical explanation of the force of gravity; he first required a sensible concept of what "motion" actually was. He had thought about this in his youth and he had of course studied the works of his predecessors. In his notebook, we can find the following statement: "Amicus Plato amicus Aristoteles magis amica veritas," which roughly translates as "Plato is my friend, Aristotle is my friend, but truth is a better friend." And Newton was right to place truth above ancient knowledge. The interpretations of the ancient philosophers' insights still formed a large part of the academic tradition in the seventeenth century, but Newton recognized that he needed to go further and to actually cast doubt on their findings. Aristotle, for instance, was convinced that an object's motion could only change if the object came into direct contact with another one that caused this change. He also believed in a "natural movement," by which he meant that some things always strived to move upward toward the sky (air, fire, smoke, etc.), while others naturally moved downward without any external impulse. The natural movements on Earth were different from those in the heavens, where natural movement always had to follow a perfectly circular path.

With the work of Copernicus, Galileo Galilei, and Johannes Kepler at the latest, people knew that the motion of the celestial bodies didn't quite work in the way the ancient Greeks had imagined. But it was still unclear in Newton's time what motion actually was and how it could be accurately explained. In 1664, one year after Newton's birth, the French mathematician and philosopher René Descartes published his major work *Principia Philosophiae*. In these *Principles of Philosophy*, he pondered the nature of all manner of things, including the motion of the celestial bodies. He developed a vortex theory, according to which the cosmos was filled with a kind of liquid celestial matter called "ether." All objects moved within this ether and their motion was caused by vortices emanating from the sun. Like Aristotle, Descartes was convinced that mechanical contact between the mover and the moved was necessary and so a vacuum could not therefore exist. The ether between the planets was needed so that the sun could drive it like a motor with its rays and thus set the entire cosmos in motion.

But Newton was not merely concerned with motion in the heavens. In his *Principia*, he wanted to answer two questions: How can motion in general be understood and explained? And how can one use this explanation to understand both the motion of the celestial bodies and that of objects on Earth? In his preface to the *Principia*, Newton makes it clear that, unlike Descartes, his aim is to grasp and explain natural phenomena using mathematical rules. He considered the Frenchman's philosophical approach to be nonsense—one might be able to use it to "make a name for yourself . . . [but it] is hardly better than a fairytale."

Newton described the aim of his own work in an astonishingly modern way: "from the phenomena of motions to investigate the forces of nature, and then from these forces to demonstrate the other phenomena." That's exactly how modern natural sciences work. One observes certain phenomena in nature and tries to deduce from them principles and theories with which one can make much more general statements and predictions about still unknown phenomena. This wish not simply to explain specifically why the moon orbits the earth, but to find a universal explanation, is made clear by Newton in another part of his introduction, when he describes what exactly he wishes to investigate: "the science of motions resulting from any forces whatsoever." He is concerned not only with gravity, but also with everything that might have any sort of influence on the motion of objects.

THE INVENTION OF A NEW LANGUAGE

In order to achieve such a general understanding of nature, Newton first needed a completely new language. Words like "force," "mass," "momentum," "space," or "time" are clearly defined in the world of science today. But Newton didn't have such definitions at his disposal, and he accordingly faced a struggle when it came to finding clear descriptions. What is a "force" in a physical sense? How can it be demonstrated and how is it measured? Using the official standard unit, the "Newton," of course, as all schoolchildren learn today in physics classes! But how did this unit come about?

In the notebooks of his youth, Newton wasn't yet decided which word he should use for which meaning: "power," "efficacy," "vigor," "strength," and other similar terms are used interchangeably. They all have connected meanings, but for a mathematical natural philosophy of the kind Newton wanted to create, he needed clear definitions, not a dictionary of synonyms.

It is precisely with a list of such definitions that Newton's *Principia* begins. He starts off with mass, defining the mass of an object as the amount of material contained within it, and also explaining that in his definition he has no need for any old "ether" or anything else which might be found between the parts out of which an object is composed. Definition number two in the *Principia* is concerned with the highly important quality of momentum. Newton calls this "quantity of motion" and explains that it is measured using the speed and size of the material (i.e., the product of the mass and velocity). Of at least equal importance for Newton's physics is the third definition. Here he explains that "inertia" (or *vis insita*, as he calls it) is the characteristic with which every object works against a change to its current situation (either at rest or in motion). And he uses this definition in order next to establish what a "force" is: namely, an action that is exerted on an object in order to change its state of motion.

If all of that sounds a bit trivial from today's perspective, it just goes to show how much Newton's work has seeped into our very consciousness. Newton had just found his language, and what he had to say with it remains unforgotten by physics even after all these centuries. In the next part of the *Principia* come the three "laws" or "axioms" that bear Newton's name today and which can be seen as the founding stone upon which modern physics was built.

In their brief and precise choice of words, the three Newtonian axioms are expressed with an almost irresistible elegance:

1. "Every body perseveres [remains] in its state of rest, or of uniform motion in a right [straight] line, unless it is compelled to change that state by forces impressed thereon."

2. "The alteration [change] of motion is ever proportional to the motive force impressed; and is made in the direction of the right line in which that force is impressed."

3. "To every action there is always opposed an equal reaction; or, the mutual actions of two bodies upon each other are always equal, and directed to contrary parts."[2]

In his first law, Newton describes a universe in which everything is either at rest or moving in a regular fashion along a straight line. Once created, everything remains as it is and nothing changes. Since our universe is obviously not made like that, however, Newton's second law explains what is required to change an object's state of motion. Each and every change is caused by a force acting. How the motion is changed depends on the direction and the strength of the force.

Newton explains that the force corresponds to the temporal change of momentum. Everybody learns the modern formulation of the second law at school as the basic equation of mechanics: force = mass times acceleration.[3] The third law may sound a little confusing at first, but it is no less important than the first two. Only with its help is it possible to explain how several masses influence each other. Newton provides a pithy explanation: "If you press a stone with your finger, the finger is also pressed by the stone." Or to express it a bit more scientifically: when a force acts between two objects, the momentum of both objects changes to the same degree. Since the momentum is the product of mass and velocity, however, the change in velocity depends on the objects' individual mass.

Newton deduced everything else from these three fundamental statements,[4] and his explanation of gravity also originates from them. The first ideas for how to explain lunar motion, about which Newton and Hooke had quarreled so bitterly, were based on a balance of forces: one force which becomes weaker with the square of the distance and emanates from the center of the earth, and a centrifugal force that acts outward and prevents the moon from falling down on the earth.

Newton's laws now enabled a more elegant explanation for the motion of the moon. The first law states that the moon actually moves through space along a straight line. This straight line is thus the "natural" state of its motion and so, if the moon doesn't follow a straight line—which is of course the case—then something must be responsible for this. This something is a force—*one* force and not two forces that need to be balanced out. The force that makes the moon deviate from the straight line and follow its curved path around the earth must originate in the earth itself; it is a so-called "centripetal force," i.e., a force that is always directed at a certain point (Newton described such a force in definition five in the *Principia*). Every object that moves along a circular path must be controlled by a centripetal force at the center of its path.

A UNIVERSAL FORCE

What Newton is telling us, therefore, is that the moon doesn't fall *to* the earth, but rather *toward* the earth. And it is doing this constantly. The moon strives to move along its straight line, but is accelerated toward the earth by the centripetal force being exerted by the earth. Not just once, but in every single moment—it falls *around the earth*.[5] There is nothing here yet about a "force of gravity," however. To begin with, Newton only showed that the change in the moon's motion along its path must be caused by a force that is inversely proportional to the square of the distance from the earth.

Only later did Newton calculate how strong the force exerted on the moon must be: the moon has to fall 15½₂ "Paris feet"[6] toward the earth each minute in order to follow the orbit that can be observed. If that is how strong this force is despite the distance of the moon, and it is inversely proportional to the square of the distance, how strong must its effect then be on objects near the earth's surface, for example a falling apple? The force must be stronger here, and so the apple must fall more quickly to the earth than the moon. Newton's answer: the said apple would fall 15½₂ feet per second, i.e., sixty times faster than the moon, which takes a minute for the same distance. That being said, science historians today are pretty sure that Newton must have played around with the figures a bit in order to get a perfect match at the end of his calculations. This changes nothing about the validity and brilliance of his theory, though it is further evidence of how far Newton was prepared to go to avoid potential criticism of his work.

The motion of the moon and the apple can therefore be explained by a single force. The force that causes the apple to fall in the direction of the earth's center is, at a greater distance, as strong as the centripetal force emanating from the earth that causes the moon to fall around the earth. Newton concluded that this centripetal force *was* the force of gravity. There is no difference and so it is pointless to differentiate them.

What is true of the earth and the moon is also true of the sun and the earth. Or the sun and comets. Or Jupiter and its moons. And thus Newton finally arrives at the peak of his findings about the motion of the celestial bodies and writes: "The force which retains the celestial bodies in their orbits has been hitherto called centripetal force; but it being now made plain that can be no other than a gravitating force, we shall hereafter call it gravity."

That's what the story of the apple and the moon is really about. Not the "discovery" that there is a force of gravity that makes things fall to the ground, and also not the fact that this falling can be mathematically explained by a force that is inversely proportional to the square of the distance. That is certainly important, but of greater importance is the fact that Newton could clearly establish that gravity is universal; that it is this single force that keeps the cosmos in motion and ensures that the stars, planets, comets, and moons move around each other.

"It remains that from the same principles I now demonstrate the frame of the System of the World," he writes. People would have imputed delusions of grandeur to any other scientist making such a declaration, but not to Isaac Newton. And in his *Principia* he did indeed do nothing less than explain the "System of the World."

THE MAN WHO DIDN'T WANT TO BE UNDERSTOOD

It is just a slight shame that Newton took such little trouble in this revolutionary work to make it comprehensible. Much to all his contemporaries' surprise (though not his own), he was in a position to explain the entire cosmos. You might have thought that he would want to share this with as many people as possible. But Newton, being Newton, had different ideas.

From the beginning, he had planned his *Principia* as a mathematical text and that's indeed what it was. Not only that, it was a text that was based on a completely new form of mathematics. In order to explain the motion of the celestial bodies and all the other dynamic phenomena, Newton required the mathematics that he had developed years before (see chapters 3 and 7), the mathematics that allowed him to explain the infinitely small and the infinitely large and to grasp changing values. This branch of mathematics, known today as differential and integral calculus, is a powerful tool without which the *Principia* could not have been composed. But when he wrote the book, Newton didn't bother sharing his new methods with the world.

Like at the beginning of his scientific career, when he had inhibitions about making his mathematical discoveries public, he was not prepared to do so now. Newton could have used his new analytical language[7] in order to derive and prove his conclusions about the universe as clearly and simply as possible. Instead, he used the existing language of geometry: the *Principia* is full of circles, intersections, lines, geometrical shapes, and other complicated and abstruse diagrams.

Mathematically speaking, Newton compressed his thought processes to an extreme degree and all too often didn't bother to write down the intermediate steps he had taken (especially when they were connected with his new mathematical method). That made it difficult or even impossible for many of his contemporaries to follow his work. The philosopher John Locke, for example, didn't even bother trying. He only read the part of the *Principia* that includes the results and for the rest he simply asked the Dutch scientist Christiaan Huygens whether Newton's mathematical deductions could be trusted.

That was precisely Newton's plan. He didn't want everybody to be able to understand what he had written because, as we already know, he was a shrinking violet and had no desire to be criticized. "He designedly made the *Principia* abstruse," wrote the clergyman and natural philosopher William Derham in a letter in 1733. He said that Newton had told him that he wanted to avoid "being baited by little Smatterers in Mathematicks" and only true mathematicians should be able to follow what he had achieved. In the introduction to the third volume of the *Principia*, Newton himself writes that he had actually intended to publish a version comprehensible to the general public, but had decided against this, because "such as had not sufficiently entered into the principles could not easily discern the strength of the consequences, nor lay aside the prejudices to which they had been many years accustomed, therefore, to prevent the disputes which might be raised upon such accounts, I chose to reduce the substance of this Book into the form of Propositions, which should be read by those only who had first made themselves masters of the principles established in the preceding Books." Newton clearly wasn't interested in the general public reading his work and wanted his audience to consist of experts. Such an attitude is perhaps understandable given his experiences, but it isn't particularly congenial, nonetheless.

The contents of the *Principia* certainly deserved a wider audience. Planets, comets, centripetal forces, momentum, and inertia might have been a bit too abstract for the average layman at the time. But Newton demonstrated in the third part of his work what his theory was capable of achieving. With it, not only the heavens but also many phenomena on Earth could be explained, and the most impressive example of this, and the one that was most entwined with people's daily lives in Britain at the time, was certainly the tides.

THE WILD NATURE OF THE TIDES

It is one of the ironies of history that it was Isaac Newton, a man who probably never saw the sea in his life—he spent his life in Woolsthorpe, Cambridge, and London, away from the long coastline of the British Isles—who finally found a practical explanation for the phenomenon of high and low tides. On the other hand—who, if not Newton, could have done so? Even though he never saw the sea, the world of mathematics in his head contained everything that was needed for such an explanation.

This was long overdue. Back in ancient Greece, unsuccessful attempts had been made to find one. One of the first (or at least one of the first we know of) to try to find the cause of the tides was the Greek geographer Pytheas, in the third century before Christ. He traveled from the Mediterranean as far as England and was subsequently convinced that the moon definitely had a role to play in the matter. But he had no idea what form this influence might have. The other Greek scholars, too, had little to contribute to clarifying the issue. In the Mediterranean, the tides are not particularly pronounced,[8] and it was difficult to obtain decent observation data. In the Middle Ages and the early modern period, there was no shortage of attempted explanations with differing degrees of absurdity. The Anglo-Saxon Benedictine monk Beda Venerabilis, born in the seventh century, also considered the moon to be responsible for the tides and thought that it "blew" on the sea and thus moved the water. In the thirteenth century, the Persian scholar al-Qazwini surmised that the sun and moon heated the water of the seas, with this leading to it spreading out at regular intervals. The (re)discovery of America at the end of the fifteenth century inspired the Italian natural scientist Julius Caesar Scaliger to believe that the water swashed back and forth between the European and the American continents. Johannes Kepler was on the right path in the seventeenth century, and supposed a force of attraction exerted by the sun and moon on the water of the oceans—but he thought along the lines of something similar to magnetism; his contemporary Galileo Galilei, on the other hand, believed this was nonsense and sought the cause of the tides in the rotation of the earth. René Descartes once again brought his vortices into the equation—but nobody could really provide a logical explanation for how the whole thing actually worked in detail.

Nobody apart from Isaac Newton, of course. He had created the tools to be able to calculate exactly how strong the moon's gravitational pull was on the earth. He also knew how to calculate the combined influence of several different forces and could show that the moon's influence alone was not enough to explain the alternating high and low tides observed. Both the moon and the sun together with their combined forces of gravity are responsible. The height of the tides depends on the relative position of these two heavenly bodies. Sometimes, their influences complement one another, sometimes they partly work against each other—but Newton was able to irrevocably prove that it was gravity we have to thank for the tides.[9]

Many readers of the *Principia* were extremely surprised by Newton's analysis of the precession[10] of the earth's axis. The phenomenon itself had been known since ancient Greek times: the direction in which the rotation axis of the earth points is not always the same, but rather changes over time. Today, its northern point is directed almost exactly toward the North Star. This was not always true, however, and will also not be the case in the future. The axis describes a small circle in the heavens, for the completion of which it requires about 25,800 years.

The first person to describe this phenomenon was the Greek astronomer Hipparchus in the second century BCE. Just like with the tides, there were numerous attempts to explain the phenomenon. Ptolemy, for instance, who was fully convinced that the earth was fixed at the center of the universe and therefore couldn't rotate around its axis, considered precession to be a slow rotation of the heavens themselves. Later astronomers, who had a heliocentric view of the world, correctly recognized that it must be connected to the rotation of the earth—but nobody knew exactly how it happened. Until Isaac Newton came along and simply demonstrated it in his *Principia*.

A few chapters earlier, he had looked into another long-standing question: What form does the earth have? Like all other learned people before and after, he knew that it was not a disk,[11] but a globe, though not a perfectly round one, as he could work out with his theory about forces and motion. The rotation of the earth around its axis and the centrifugal force resulting from this lead to (put very simplistically) material being pressed "outwards" toward the equator. If we measure the radius of the earth from its core to one of the poles, the distance is slightly less than if we measure it to the equator.[12] The earth bulges out a little at the equator, and this thickening creates precession: because the gravitational fields of the sun and moon have a greater surface area on which to work, this makes the earth wobble while rotating. This is precisely what Newton was able to work out and, at the end of his analysis, he arrived at a value for the speed of the precession that more or less matched what was observed.[13]

The extracts about the earth's form, the tides, and the precession of the earth's axis are perhaps the most astonishing parts of the entire *Principia*. The starting point for the work on the book was the question about the form of the orbital paths of comets. In the course of his answer, Newton not only created a completely new theoretical basis for explaining forces and motion and demonstrated that the force that keeps heavenly bodies moving along their paths is a universal one; he was also able to provide concrete evidence that the sun and the moon affected the phenomena of the earth itself with the same force of gravity. Universal gravity is much more than just a simple formula to calculate orbital paths. It is what gives the universe its form and dominates the entire cosmos in every aspect.

The *Principia* is one of the great masterpieces of humanity, comparable with the Mona Lisa, Beethoven's Ninth Symphony, or the Pyramids. Yet in contrast to art and culture, Newton's system of the world and what his successors constructed upon it are still understood and appreciated by far too few people.

BETTER TRANSPARENT THAN SORRY

Isaac Newton was not in the least concerned to make his research accessible to his colleagues or the general public. And I fear that he wouldn't have any problems getting away with this in the world of research today. Like back then, public relations play far too minor a role in science today.

Which is astonishing: of course research is still the main thing in science, but unlike in the seventeenth century, we live in a world today that is very much shaped by the insights and results of scientific research. What Newton and his peers found out was certainly revolutionary, but generally speaking it had nothing to do with most people's day-to-day lives. Today, things are different. Biogenetics, stem cell research, renewable energy, pharmaceutics, climate change, nanotechnology; not only is research carried out in all these fields—they also have a direct influence on our daily lives. Politics and society have to make the appropriate decisions and decide where and how the new knowledge gained can or may be used. That can only seriously happen if society, at least to an extent, understands these issues. Public relations in science is therefore indispensable, and not only in the context of applied research. Providing information about (seemingly) abstract basic research is also crucial if we do not wish to live in a world that is becoming more and more incomprehensible for more and more people. The computers that we use every day are based on the findings of quantum mechanics. The algorithms that show us personalized advertising on the internet and, without our noticing, evaluate our personal data, have their origins in abstract mathematics that could also come from Isaac Newton himself. And so on and so forth: if we want to have a tiny bit of control over the world in which we live, we have to know at least a little about science.

It would also be in the interests of scientists themselves to engage in as wide a range of public relations work as possible. Most of them work at state universities or research institutes and are funded by taxpayers' money. The public therefore has the right to receive information about the use of its money. But even if one doesn't care about that, one should care about the people who distribute the money. If the politicians have to save money, they will do so where they can expect least resistance from the general population. The more and the better the public is informed about the reasons and necessity for research, the simpler it is for scientists to get funding for their work.

In Isaac Newton's day, science was still to a large extent a private hobby for well-to-do men and they could afford to ignore the public. But today, it should actually be taken as self-evident that the findings of science should be communicated to as many people as possible. Nevertheless, the idea of the "ivory tower" in which scientists carry out their work, cut off from the rest of the world, is unfortunately still somewhat more than just a cliché.

If you want to have a successful career in science, there is still one thing above all that you need to do: publish scientific papers. The range and quality of your publication list are the number 1 assessment criterion when it comes to getting a job or receiving funding for a research project. If you spend too much time on public relations, then you will be displaced by those who don't bother with that and instead prefer to publish a few articles.

Somebody like Isaac Newton, therefore, who concentrated exclusively on research and had no interest in interaction with the public, would still get along very well in the world of science today. But I still wouldn't recommend basing oneself too much upon him in this regard. There is still the hope that the current situation will change some time. Slowly but surely, the scientific funding organizations and big research institutes seem to understand that it is worth communicating directly with the public. In the USA, for example, it is already almost a matter of course that every major NASA space mission is accompanied by special social media appearances to provide information to the public. In Germany, too, young scientists address the public in YouTube videos, internet blogs and podcasts, or appear at Science Slams in order to present their research in a clear and fun way. Sooner or later, the value of science communication will find its way into the structure of the universities and then it will hopefully be beneficial rather than detrimental to one's career to devote oneself to public relations.

CHAPTER 6

IN
SEARCH
OF THE
PHILOSOPHER'S
STONE:
NEWTON'S
ESOTERIC SIDE

A few days before his death on March 31, 1727, Isaac Newton was still working on a manuscript that he hoped soon to publish. The text had nothing to do with mathematics, physics, or astronomy, however. It contained no new scientific discoveries and no further revolutionary ideas that would occupy natural scientists until the present day. The book was called *The Chronology of Ancient Kingdoms* and was Newton's attempt to sort the history of the world and, above all, the stories of the Bible, and to date them. He looked into the geometry of Solomon's temple as described in the Old Testament, for instance, seeking in the building's measurements given in the Bible hidden messages and secret wisdom.

The search for secret knowledge in old religious texts seems like the sort of thing for pseudoscientists like Erich von Däniken and certainly not the man whom we view today as the pioneer of modern natural sciences. But *The Chronology of Ancient Kingdoms* is strictly speaking more typical of Newton's work than his scientific texts. Newton owned a total of thirty Bibles—but only thirty-three books about astronomy. In his private library, there were almost five hundred books on theology, but only fifty-two books about physics. When (most of) the papers from his estate were auctioned off in 1936, the catalogue comprised texts with more than three million words written by him. Almost half of these are concerned with theology, religion, and the interpretation of the Bible. He devoted approximately 650,000 words to alchemy.

Newton was a great natural scientist, but of equal or perhaps even greater importance to him was his work on subjects that, from today's perspective, could hardly be further removed from physics, mathematics, or astronomy. Bible interpretation, religious chronology, mystical alchemy, and prophecies: this esoteric side to Newton was long ignored and he was instead, like in the preceding chapters of this book, presented as the great rational genius whose outstanding intellect deciphered the concealed laws of nature, as a calculating thinker whose discoveries in mathematics and physics prepared the way for the scientific conquest of the world for the following generations, as the pioneer of a mechanistic view of the world that has served as the basis for modern science until today. And yet he was also:

THE LAST OF THE MAGICIANS

The British economist John Maynard Keynes, an enthusiastic collector of old books, purchased a large selection of Newton's papers when they came up for auction. From him come these words that beautifully sum up Newton's approach: "Newton was not the first of the age of reason. He was the last of the magicians."[1] Newton founded the modern scientific era, but he himself was not a part of it. It was just as important to him to concern himself with religion as with optics, mathematics, or the motion of the celestial bodies. And just like with his scientific research, he paid no heed to conventions or authority in his theological studies. It wasn't merely Newton's preoccupation with religion itself that ensured that his texts about God and the Bible remained unpublished for so long. It was above all their contents, which were in opposition to the church's doctrine. Newton didn't merely study theology. He was also a heretic.

Even when he was on his deathbed in March 1727, he refused the last rites of the church. And long before that, in 1675, he was on the point of resigning from his position at the University of Cambridge. As was customary for those working at the university at the time, he was supposed to be ordained in the Anglican clergy. Before he could begin his studies in Cambridge, he had to swear to remain celibate and to accept and keep to the statement of faith of the Anglican Church. However, he was not prepared to take the next step and become ordained in the clergy itself.[2] He believed in God, but not in the way prescribed by the church.

Newton was convinced that the religious scriptures had been corrupted over the course of time and that what the church taught at the time was not the original message. He believed that people in the past had still been aware of the true word of God, and only by precisely analyzing the original words could the rest of the truth be found and reconstructed. He undertook his comprehensive studies partly in order to prove this. And what he considered to be the original message was considered dangerous. The Blasphemy Act of the English Parliament in 1697 expressly forbade the denial of the Holy Trinity, with punishment for this ranging from the removal of public offices to several years in prison. But Newton believed precisely that—that the concept of the Trinity was wrong. For him, there was just one God, not "the Father, the Son, and the Holy Spirit," Jesus had indeed lived, and he had been the Son of God. But indeed, only the Son.

Twenty-seven years after Newton's death, *An Historical Account of Two Notable Corruptions of Scripture* was published, a book with texts that he had written back in 1690. In it, he analyzes two passages from the Bible that describe the Trinity of the Christian God and attempts to use the comparison of older editions of the Bible in various languages to show that these passages do not appear in the original Greek text of the New Testament. Newton was concerned with this sentence from the First Epistle of John: "For there are three that bear witness in heaven: the Father, the Word, and the Spirit, and these three are one. And there are three that bear witness on earth: the Spirit and the water and blood, and these three agree." He believed it was, at least in part, a later addition and could not be found in the old Greek manuscripts, and analysis in modern religious studies proves him right. In most current Bible editions, most of this part, known today as the Comma Johanneum, is left out or only included as a footnote.

In order to prove that the Trinity doctrine was wrong, Newton didn't merely analyze ancient religious texts—he also used mathematical and geometrical arguments. Having calculated in astronomy the forces between the celestial bodies, he used similar diagrams here to show that, while Jesus was the Son of God, the divine power could only come from God Himself, and not from Jesus. From today's perspective, these endeavors seem naive, of course. Newton drew a picture, for example, consisting of a rectangle with three strips (or three rectangles on top of each other). According to him, the image was supposed to represent three "bodies" (A, B, and C) lying and pressing on top of each other. The top one (A) exerted a force (of gravity); the two bottom ones didn't. But since the top one pressed down on the other two, they too would feel a force. Or, as Newton expressed this in an unnecessarily complicated way: "There is then a force in A, a force in B and the same force in C. But these are not three separate forces but ONE force originally in A, and by communication and descent, in B and C." That was supposed to be the proof that there is only one God with a divine power and not a Holy Trinity consisting of God the Father, Jesus Christ, and the Holy Spirit. Apparently, even geniuses can sometimes be truly simplistic.

Newton's theological activities went far beyond an attempt to disprove the doctrine of the Trinity, however. Just as he had come up with a universal explanation for a number of phenomena with his *Principia*, he also saw the ancient religious scriptures as a universal source of information. The interpretation of the Bible was not for him unprofitable conjecture, but rather a matter of supreme importance. He was convinced that the true word of God was to be found there and that God, unbound by space or time, had set out man's future there, though in a coded language that Newton attempted to decipher. The prophetic texts in the Bible were for Newton "histories of things to come."

Newton's works on theology and natural sciences were for him two sides of the same coin. Just as he could investigate God's creation by studying nature, he could also use the religious scriptures for the same purpose. For this, though, he needed to find out their original meaning and couldn't rely on what he considered to be the corrupt doctrines of the church. You can see how deeply he immersed himself in the world of the Bible from the work *Observations upon the Prophecies of Daniel and the Apocalypse of St. John*, which was also only published posthumously in 1733. The titles of the individual chapters form a stark contrast to his texts about physics and astronomy. Chapter 8, for example, is called "Of the power of the eleventh horn of Daniel's fourth Beast, to change times and laws." The Bible passage from the book of Daniel, which Newton goes into here in great depth, could hardly be more different from the physics that one normally associates with him: "After that, in my vision at night I looked, and there before me was a fourth beast—terrifying and frightening and very powerful. It had large iron teeth; it crushed and devoured its victims and trampled underfoot whatever was left. It was different from all the former beasts, and it had ten horns. While I was thinking about the horns, there before me was another horn, a little one, which came up among them; and three of the first horns were uprooted before it. This horn had eyes like the eyes of a human being and a mouth that spoke boastfully." (Daniel 7:7–8)

A PROPHET OF DOOM AND AN ALCHEMIST

Newton spent a large proportion of his time with such abstruse, mystical texts. But he was convinced that, in the confusing world of the biblical prophets, just as many valuable secrets about the universe could be found as in the (at that time equally confusing) natural world.

The wider public first became aware of Newton's preoccupation with the Bible in 2003. That year, the BBC showed a documentary providing an in-depth look at the great physicist's theological research, a subject that had until then been of interest to just a small number of science historians. And this probably would have still been the case, if the documentary hadn't claimed that Isaac Newton had predicted that the world would end in 2060.

Isaac Newton, the great scientist, a prophet of doom?

Yes—and no. In his texts, this date can indeed be found. Due to his studies of the biblical scriptures, Newton was convinced that the "Kingdom of God" would one day prevail on the earth. First, though, the age of the corrupt church must come to an end. For Newton, this age of course began with the introduction of the false doctrine of the Trinity of God, which Newton dated to the symbolic year of 800, when, in his view, the hegemony of the popes began. From the nebulous texts of the biblical prophets, he derived a period of 1,260 years, which had to pass before the false doctrines of the corrupt church would disappear once again, and thus he arrived at the aforementioned year of 2060—which for him signaled not the apocalyptic end of the world, but rather a spiritual new beginning. He even spoke out in no uncertain terms against the foretellers of doom who were active at the time and who struck fear into people's hearts with their visions of the pending end of the world. After working out the year 2060, Newton wrote: "It may end later, but I see no reason for its ending sooner. This I mention not to assert when the time of the end shall be, but to put a stop to the rash conjectures of fanciful men who are frequently predicting the time of the end, and by doing so bring the sacred prophesies into discredit as often as their predictions fail."[3]

All of this simply doesn't tally with the common perception of Newton. But while his profound belief and preoccupation with religion are certainly understandable within the context of seventeenth-century society and circumstances, this is not so much the case with another of his interests. He was not only a great scientist and a heretical theologian—he was also a staunch alchemist in search of the philosopher's stone.

"The staff of Mercury unites the two serpents and entwines them through the connection of Venus.... This red powder is accordingly Flamel's male wingless dragon, for after it has been extracted from its normal powder, it is one of the three substances out of which the bath of sun and moon is made." One might expect to find such words in muddled texts by would-be medieval magicians. Yet they are by Newton. He did indeed write about "fiery dragons" and "the Blood of the Green Lion" and "the stone of the ancients" and was an alchemist through and through—perhaps even more than he was a natural scientist.

Especially since these two occupations did not, in his view, contradict one another. Like his interest in religious scriptures, alchemy was also just one aspect of a comprehensive search for insight. The "first religion" contained the truth, while the "ancient ones" had the key to the truth and had concealed this knowledge in a cryptic form in ancient writings. For Newton, religion, alchemy, and nature were closely linked to one another. "The true God ... is Eternal and Infinite, Omnipotent and Omniscient," he wrote, adding, "He is omnipresent, not virtually only, but also substantially." Newton was in search of God and the way in which He influenced the world. He once conjectured that there must be a kind of "divine matter," a substance fundamental to the matter out of which everything is made and with which everything can be transformed. His aim was to provide experiment-based evidence of the divine intervention in the transformation of matter.

Alchemy occupied Isaac Newton for almost his entire life. When he lived with the apothecary William Clarke during his school days in Grantham, he came into contact with the world of chemistry and alchemy. The first experiments of his own were undertaken in 1668 during his time at Cambridge University, where he also mixed his own medicine to protect himself against all possible diseases: a potion made of turpentine, rose water, olive oil, beeswax, and liqueur.

He set about actual alchemy with the same dedication as he did the rest of his undertakings. He constructed his own furnaces and chimneys and began working on a comprehensive lexicon of alchemical terms, the "Index Chemicus" that would finally contain more than nine hundred entries. He recorded his experiments with the utmost precision, and the sentences about serpents, dragons, and green lions quoted above come from these records. Did poisonous vapors from the chemical laboratory cloud Newton's mind here? No—there is another explanation for these esoteric-sounding terms.

Alchemy was already on the wane in Newton's day. In the Middle Ages, the search for the "philosopher's stone," which was supposed to turn lead into gold and ensure eternal life, had been much more widespread. Alchemists themselves, however, were not always regarded particularly favorably. Among those who were seriously engaged in studying the characteristics of the chemical elements could always be found charlatans and swindlers aiming to trick gullible adepts out of their money. And those in power were also not necessarily keen for people to be able to produce money at will (except for when the alchemists were in their service). In England, therefore, the "Act Against Multiplication" was made law by King Henry IV, making it a punishable offense to "multiply gold and silver."

In this case, therefore, Newton actually had a good reason for once to keep his experiments secret. But he did so not only for fear of prosecution or the loss of his reputation as a serious scientist. Ethical and philosophical reasons were even more important for him. From his point of view, alchemy was a spiritual undertaking—the search for God and the fundamental secrets of creation. Neither the preoccupation with metals nor the production of gold was the true aim of alchemy, but rather "to glorify God in his wonderful works, & to teach a man how to live well." Not everybody could do this, Newton explained, since "They who search after the Philosopher's Stone by their own rules [are] obliged to a strict and religious life," and the metaphysical use of furnaces and crucibles was not something that could be disclosed to the ordinary rabble.

For that reason, Newton wasn't in the least pleased when

his colleague Robert Boyle, in February 1675, published an article titled "Of the Incalescence of Quicksilver with Gold" in the Royal Society's *Philosophical Transactions*. Boyle, fifteen years Newton's senior, was one of the founding members of the Royal Society. He concerned himself, among other things, with the characteristics of a vacuum and with chemistry. His most famous work is *The Sceptical Chymist*, published in 1661, in which he endeavors to distinguish scientific chemistry from the not-so-scientific alchemy. And although he is viewed today as the founder of modern analytical chemistry, his thinking was shot through with alchemistic ideas. He too believed in an "original matter," a "philosopher's stone," and that chemical elements could be transformed into one another, if one only knew the right recipes.

It was therefore only logical that Boyle should campaign for the abolition of the Act Against Multiplication, which he managed to achieve in 1689. He was convinced that the scientific study of the characteristics of material was possible and worthwhile, and had no problem with publishing his research findings. Newton, on the other hand, was enraged that an alchemistic colleague not only went public, but also did so in language that could be generally understood.

The stories of serpents and dragons in Newton's records were not a mystical end in themselves. They were designed to make the experiments and recipes appear confused and incomprehensible, so that only the truly initiated would know what they were all about. If you wanted to say, for example, that you should mix two substances while applying heat, the instruction was to "send the two dragons into battle." There were "male" and "female" elements, and when two of these coalesced during a reaction, then this was a "marriage."

ON THE TRAIL OF THE PHILOSOPHER'S STONE

Those who weren't prepared to master this complicated language were also not worthy of setting off in search of God and the philosopher's stone. The alchemists also communicated with one another using pseudonyms, rather than their real names. Newton, for instance, called himself (with the utmost modesty) "Jehova Sanctus Unus," a holy god.[4]

While Newton's work formed the basis for a mechanistic view of the world that would dominate the following centuries, Newton himself believed in a quite different world. The entire cosmos and all of matter were in his view pervaded with a divine spirit, and it was important to recognize and grasp this. Alchemy was, therefore, not merely a hobby or—as has often been said—something that he only started doing late in his career, when his mind was not as sharp as before. If anything, it would be closer to the truth to call Newton's research into physics a "hobby" that he fitted in between his theological and alchemistic studies. When Edmond Halley spoke to Newton about the question of the mathematical explanation of comets' orbits and thereby prompted him to write the *Principia*, Newton was right in the middle of alchemistic experiments. In the preceding years, he had hardly carried out any research into physics and had instead been almost exclusively engaged in trying to create the philosopher's stone in his laboratory. And after the publication of the *Principia*, too, he completely devoted himself once more to alchemy.

In 1692, Robert Boyle died, and Newton discussed what he had left behind in correspondence with the philosopher John Locke. He was interested in a mysterious "red earth" that Boyle was said to have produced in his experiments. Newton was cautious and didn't really want to address the matter directly with Locke, who first had to prove to him that he was an "initiate." Only when Locke showed that he knew what the "red earth" was about and promised to keep the whole thing secret did Newton feel safe enough to correspond extensively with him. He explained that he wanted to test his alchemistic knowledge in order to find a special quicksilver[5] that might possibly enable the production of gold.

By 1693, Newton had sent enough dragons into battle against each other, united enough serpents in matrimony, and bled enough green lions. He summarized his alchemistic research in a text of some 5,500 words and announced that "amalgaming the stone with the mercury of 3 or more eagles and adding their weight of water, & if you designe it for metals you may melt every time 3 parts of gold with one of the stone. Every multiplication will encreas it's virtue ten times &, if you use the mercury of the 2nd or 3rd rotation without the spirit, perhaps a thousand times. Thus you may multiply to infinity."[6] Newton was convinced that he had found a "stone" with which gold could be multiplied "to infinity." After his break-throughs in physics and mathematics, he had now achieved his spiritual and alchemistic goal: he had discovered the philosopher's stone.

But though Newton was often right, in this case he was wrong. The philosopher's stone remained out of his reach. Later, he never returned to his supposed discovery. After the seemingly triumphant conclusion of his alchemistic research, he suffered a nervous collapse, the cause of which is still unclear today.

In September 1693, Newton wrote a confused letter to Samuel Pepys, who had been the president of the Royal Society between 1684 and 1686, but had held no official post since 1690. In it, he wrote of an "embroilment" in which he was involved and which was causing him great difficulties. He said he could never again see Pepys or "the rest of my friends." John Locke also received a letter. Locke, according to Newton, had endeavored to bring him into contact with women, which had deeply distressed him, to the extent that, in a conversation with somebody else, he had wished death upon Locke, for which he now wanted to apologize. Something wasn't quite right with Newton. He broke off his research, wrote practically no more letters, and behaved in a highly curious fashion. When later asked, he explained that he had suffered massively from insomnia, which had kept him awake for up to five nights in a row. In addition, there had been a "distemper," which had caused him difficulties.

What really happened can no longer be reconstructed today. Perhaps it was indeed the recognition that his great breakthrough in the search for the divine spark of matter had failed that caused him to break down. Perhaps it was a fire in his laboratory, which destroyed many of his records.[7] More speculative minds even suggest love problems, supposedly connected with the rupture of his (epistolary) friendship with the Swiss mathematician Nicolas Fatio de Duillier. Newton's sexuality has always been one of the subjects about which there is much speculation, but little concrete information. He never married and is said to have confessed on his deathbed to one of his doctors that he had never had sex and would die a virgin. There is no evidence of close relationships with women; his intense friendship with Fatio de Duillier is documented, however, and contact between the two of them did indeed break off—also for unknown reasons—just at the time when Newton had his breakdown. Was Newton therefore perhaps homosexual and in love with Fatio, and was the nervous collapse simply due to a broken heart? We do not know. Perhaps the quicksilver vapors that Newton constantly breathed in during his experiments poisoned him. Perhaps it was depression, perhaps burnout, perhaps the pressure that he felt after the publication of his *Principia*. Whatever it was that was responsible, the crisis soon passed and Newton was ready for new challenges.

He found these soon afterward as the warden of the Royal Mint, as already described in chapter 1. And of course in his favorite occupation—quarrelling with his peers. In this respect, the quarrels he had had before had been nothing more than a warm-up. The biggest, longest, and most intense row of his career was still before him.

NATURAL SCIENCES WITHOUT GOD

Isaac Newton's religious and alchemistic studies would most likely elicit much more than just a raised eyebrow from his colleagues in the world of science today. It's almost impossible to imagine a modern scientist combining serious science with apocalyptic prophecies and an esoteric search for the philosopher's stone. This is above all due to the fact that we now know much more about the world than people in Newton's time did. In the seventeenth century, nobody knew how matter was really constructed. Nobody could have known, because the technical instruments required were a long way from being invented. All natural philosophers had back then was speculation, and the chemical knowledge of the early modern period made it seem a quite plausible idea that it should be possible to transform chemical elements into each other if one could only find the right mixture.

Religion, too, had a completely different standing. Atheism was simply not an option. And the existence of God or a "higher being" was the practically self-evident basis for any search for knowledge. The fact that Newton kept his theological investigations secret was due to the fact that the belief he was thereby attempting to prove contradicted the doctrines of the church, and not to the fear of appearing not to be a legitimate physicist because of his belief in God.

Today, things are very different. For phenomena where "God" was previously cited as the reason, we have now found better and more scientific explanations. There is no need to conceal your atheism, and many scientists don't believe in God. On the other hand, many of them do—and that is also absolutely no problem. It is possible today to be just as religious as in the seventeenth century without having to fear any negative consequences for your career. With one qualification: you should definitely avoid mixing your personal beliefs with science.

For Newton, however, there wasn't such a separation. In his *Principia*, he even specifically pointed this out: "This most elegant system of the sun, planets, and comets could not have arisen without the design and dominion of an intelligent and powerful being . . . . This being governs all things, not as the soul of the world, but as Lord over all: And on account of his dominion he is wont to be called Lord God *Pantokrator*, or Universal Ruler."[8] He was convinced that, while the motion of the celestial bodies obeyed his law of universal gravity, a higher being was nonetheless required to control everything. God always had to make sure that everything was in order, so that the edifice of the universe could continue to revolve.

Today, knowledge and belief have long since been separated, and natural sciences can no longer be based on phenomena which you have to believe in and for which there is no concrete evidence. And if this does happen, then—unsurprisingly—problems arise.

In January 2016, for example, Chinese researchers published a paper about the anatomy of the human hand in the *Public Library of Science* journal (*PLOS ONE*).[9] Here, the authors write that the human hand possesses a "proper design by the Creator to perform daily tasks in a comfortable manner." In other parts of their article, they also made reference to a "Creator" responsible for the characteristics of the human hand. As is usual in the publication of scientific papers, the article was checked by peer reviewers, but obviously not with particular care. At least they didn't seem to notice the references to a creator—but plenty of other scientists did. *PLOS ONE* was deluged with criticism, and the article in question was swiftly withdrawn. And yet the Chinese authors most likely had no hidden agenda; they themselves subsequently said that the incident was the result of their poor English and a bad translation. By "Creator," they hadn't meant a divine being, but simply nature itself, in the evolutionary context of which the human hand has developed.

Be that as it may, whether you wish to see that as an excuse or a plausible explanation, this episode shows that science today is something that you shouldn't need to believe in. Either it is possible to prove a claim with concrete measurements or observations or it isn't—in which case the whole thing has to be classed as an assumption or a hypothesis, and you mustn't fall into the trap of viewing your own beliefs as a generally binding truth, regardless of what those beliefs are.

The vast majority of Isaac Newton's scientific findings have survived until today. They need no religious justification; it is sufficient to observe nature to receive proof of their validity. The gaps in them, which he attempted to explain with God, have now been filled by verifiable knowledge. Not completely, perhaps, but no serious scientist today would consider filling them with religious speculation and presenting this as research. And that's how it should be. Isaac Newton showed us that we can understand the world; that we can explain it in a mathematical and objective way and find general natural laws. He himself probably wouldn't have been particularly pleased by this, but he showed us how to separate belief from knowledge.

Sixty-nine years after Newton's death and just before the end of the eighteenth century, the French mathematician Pierre Simon Laplace published his famous *Exposition du système du monde* (*The System of the World*). Here, Laplace concerned himself with the same questions as Newton had done in his "System of the World," particularly that concerning the motion of the celestial bodies. Thanks to new mathematical methods, Laplace had a much clearer understanding than Newton how this motion worked, and the majority of modern celestial mechanics (the science of the motion of celestial bodies) is still based on Laplace's work. There is a well-known anecdote about a conversation between Laplace and Napoleon Bonaparte, who had read the *Exposition du système du monde* and said to the author: "Newton spoke about God in his book. I have looked through yours and couldn't find a single instance of this term." To which Laplace laconically replied: "I had no need of that hypothesis."

It is highly likely that such a conversation never took place like that; at least there is no reliable evidence for it. But the anecdote was already doing the rounds at the beginning of the nineteenth century, which shows how circumstances had changed. What had earlier remained unexplained and had thus been ascribed to God's influence was now understood without God.

It was the same with alchemy, except that things developed here a little more clearly and with much less controversy.

Though Isaac Newton was an alchemist through and through, the peak of alchemistic research had already passed in his day. The constant new discoveries about nature quickly led to a demarcation between serious, scientific chemistry and pseudoscientific alchemy, and the same process took place in other fields of research. Astronomy was separated from astrology, serious medicine from quackery, and so on. A chemist who practices alchemy, or an astronomer who compiles horoscopes, is unimaginable today, even though this was completely normal in Newton's day. That is precisely the problem, however, when we wish to learn lessons for the present day from the concept of science of that time: with the knowledge we have today, it is easy to criticize Newton for his unscientific, alchemistic studies (or Johannes Kepler for his astrological work), but that only works in retrospect. Who knows what people in a few hundred years will think of certain fields of science today? Perhaps some of what we consider today to be serious research will later be recognized as being as nonsensical as Newton's alchemy.

As far as the religious, esoteric, and pseudoscientific aspects of his work are concerned, Isaac Newton is certainly not a good role model from today's perspective. Everybody can think in private what they want, but scientific work should be kept strictly separate from personal beliefs. That doesn't mean, however, that you can't be inspired by your completely private and personal world of ideas. Richard Westfall, the science historian and author of the Newton biography *Never at Rest*, which is still considered to be the definitive reference work today, is of the opinion that "the Newtonian concept of force embodies the enduring influence of alchemy upon his scientific thought" and that the alchemical tradition was the source of Newton's concept of attraction.[10] This view is not without controversy, but Newton himself certainly saw no qualitative separation between physics and alchemy. He saw "invisible" forces of attraction not only between the planets in the heavens, but also in the crucible of his alchemistic experiments. In a text about acids ("De natura Acidorum"), he writes of "acid particles" that are "endowed with a great attractive force," and in the *Principia*, he speculates about the existence of many other (attractive) forces that act between particles of matter and are yet to be discovered.

Science is a highly creative undertaking, and a great deal of imagination and fantasy is required to try to solve the mysteries of nature. Belief, myths, or other "dubious" ideas are as valid as anything else as a source of inspiration, and there are plenty of examples of this in the history of science. In 1890, chemist August Kekulé recounted how he had only discovered the structural formula for benzene because, during a daydream, he had seen a snake biting its own tail, which he then later interpreted as the molecular bond between atoms. The famous mathematician Srinivasa Ramanujan, who worked with Godfrey Hardy at the University of Cambridge at the beginning of the twentieth century, allowed himself to be guided in his dreams by the Hindu god Namagiri.

What Isaac Newton's physics and mathematics would have been like had they not been inspired by religion and alchemy cannot be said. One thing is certain, however: we only continue to use his findings in natural science today because, irrespective of their possible theological or esoteric origins, they have been scientifically confirmed in a legitimate, comprehensible, and objective fashion. And that is what matters all the more today.

That is also the reason why modern natural sciences are dominated by mathematics to such a degree. Mathematics provides a language in which we can speak rationally about nature, with which clear and verifiable predictions can be made, and which allows us to describe phenomena, which often appear confusing at first sight, in an orderly and objective way. Isaac Newton was instrumental in formulating this language, and it wouldn't have been Isaac Newton if a long and fierce confrontation had not erupted over it.

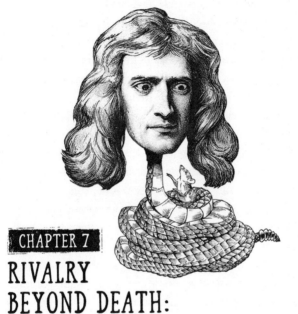

CHAPTER 7

RIVALRY
BEYOND DEATH:
NEWTON AND HIS INTRIGUES

Isaac Newton ended his life as the warden of the Royal Mint, and his struggle against the counterfeiters and the rescue of British coinage left him hardly any leisure for new scientific research. But it seems that Newton could always find time for a proper quarrel, and in the last part of his life, he finally found a worthy opponent. His row with the German polymath Gottfried Wilhelm Leibniz is the stuff of legend and has kept science historians busy for centuries. It was a battle between giants; Leibniz was not only as fanatic a scientist as Newton, and driven by the same wide-ranging thirst for knowledge, but he could also be equally relentless in imposing his will. The reason for the dispute was befitting of the adversaries: it concerned the invention of a completely new type of mathematics, the significance of which for the future development of science cannot possibly be overstated.

The question which forms the basis of this mathematics, however, is astonishingly simple and, at first sight, completely harmless: How do you demonstrate change? It's easy to see why this question was of interest to Newton. After all, motion is nothing more than a change of location, acceleration is a change of speed, force a change of momentum, and he had thought deeply about all of these concepts in order to be able to write his *Principia*.

HOW DO YOU DESCRIBE MOTION?

Newton was not the first person to wrestle with this issue, however. The question about the nature of change had been discussed more than two thousand years before his birth in ancient Greece. In the fifth century BCE, the philosopher Zeno of Elea thought up his story of Achilles and the tortoise. He described a race between the famous hero and the notoriously slow animal. The tortoise gets a head start of a hundred meters, and then the race begins. Achilles sprints off and quickly runs the first hundred meters. In this time, however, the tortoise has also advanced a little bit, so Achilles is not yet level. Naturally, he easily runs this small distance, but still needs time to do so and during this time, the tortoise advances another little bit. Its lead does get constantly smaller, but—according to Zeno—Achilles can nonetheless never overtake it. In the time that he requires to cover the distance to the tortoise, it always manages to advance a little bit further and so he cannot win the race.

Zeno had a soft spot for such paradoxes. In another well-known case, he explained that an arrow shot from a bow cannot actually move in reality. If one observes the arrow during its flight at a particular point in time, it also occupies a very specific place in the air. It must be at rest in this place, otherwise it wouldn't be in a place. What is valid for one point in time must be valid for all points in time during the flight, which must mean that the arrow is constantly at rest and so motion is impossible. Or, as Zeno himself nicely put it: "What is in motion moves neither in the place it is nor in one in which it is not."[1]

Now, Zeno wasn't an idiot and he must surely have known that something about his arguments wasn't quite right. He hadn't come up with them in order to actually prove the impossibility of motion, and it was clear to him that they contradicted people's experience. What he probably wanted to do was to defend his teacher Parmenides. Parmenides was of the opinion that—to put it simply—everything merely exists and there can be no "becoming" or "passing away," since that would mean that "something" would have to arise out of "nothing." This idea is, of course, open to criticism—but Zeno wanted to demonstrate that the arguments of his opponents, who assumed the reality of change, could also lead to contradictions. He couldn't yet recognize the contradictions that he created himself with his reasoning. Zeno believed that an infinite series of additions must give an infinitely large result. He also thought that, by dividing the running distance an infinite number of times, this would mean that the distance to run must be infinitely large. This was wrong, but the world had to wait for Isaac Newton and Gottfried Wilhelm Leibniz to come along, in order to be able to understand this with mathematical precision.[2]

Whatever one thinks of these arguments among the ancient Greeks—the question of the nature, and above all a mathematical demonstration, of change and motion remained unanswered for a long time. And it was a fundamental issue that urgently needed to be solved if you wanted to make sensible statements about moving objects.

If I set off on my morning run through Paradise Park in Jena, for instance, and cover a distance of ten kilometers in fifty minutes (thanks for your admiration; not bad, I know), it's not difficult to calculate that I run at an average speed of twelve kilometers per hour. But I certainly wasn't running at a constant twelve kilometers per hour for the whole fifty minutes. Just over a third of the way in, for example, there is a place that is a bit downhill and I generally run this part a bit more quickly, probably at about fifteen kilometers per hour. But in the last part of my jogging route, there are lots of traffic lights and crossings that slow me down, and I am certainly much slower than twelve kilometers per hour there.

It is also not difficult to calculate appropriate average speeds for each individual kilometer of my run. Or for all one-hundred-meter intervals. Or even for every individual meter, centimeter, or millimeter. Nevertheless, these are still average values that tell me how quickly I completed a certain distance. But what if I want to know how quick I am at a very specific moment? Then we are back with Zeno's paradoxical story of the flying arrow. In a single moment, I do not cover any distance; there is no line along which I have moved at a concrete speed.[3]

From a mathematical point of view, this leads us to the

question of infinity. We can make the distance for which we want to calculate an average speed smaller and smaller and simply need to divide this distance by the time it takes us to cover it. As long as the length of the distance is not zero, that's a simple task. But at some point between "zero" and "arbitrarily small," we meet infinity—and this was mathematically difficult to grasp in the seventeenth century. In order to calculate the speed of an object during a very specific moment, the distance has to be made infinitely small. But how?

It is worth also briefly viewing the problem from a geometrical point of view. Both Newton and Leibniz were influenced in their mathematical work by René Descartes. Among the latter's great achievements was the combination of geometry and algebra, that is to say the connection between mathematical diagrams and mathematical equations. Descartes showed that geometrical problems can also be formulated as equations—and vice versa—and he contributed to the resolution of the "tangent line problem" that is a central part of the new mathematics of Newton and Leibniz.

It is simple to draw a diagram that represents a movement. In the example of my jogging route, you would plot the time on the x axis and the place where I am at each moment in time on the y axis. At the point in time zero, I have covered zero kilometers: the starting point lies at the coordinates 0 min. / 0 km. Five minutes later, I have reached a kilometer further, and the next point in the diagram is at the coordinates 5 mins. / 1 km. Then perhaps I run a bit quicker and reach the two-kilometer mark after nine minutes and so enter a further point at the coordinates 9 mins. / 2 km. The next point follows perhaps at 14 mins. / 3 km, and so on and so forth: I can enter as many points with values for kilometers and minutes as I like and will end up with a graph that illustrates my movement during my run. In order to calculate the speed, you simply have to compare two selected points in time. If I want to know my average speed for the second kilometer, for example, I take the corresponding point with the coordinates 9/2 and subtract the point one kilometer before—in this case 5/1. The calculation is simple: $9-5 = 4$ minutes, and $2-1 = 1$ kilometer. So my average speed during the second kilometer was 4 minutes per kilometer, which works out to be 15 km/h. The problem can also be solved geometrically, by simply drawing a straight line through the points 9/2 and 5/1. The steeper the line, the quicker I was. If it is horizontal on the diagram (so parallel to the x axis which shows the time), the speed was zero. The time coordinate has changed, but the place coordinate hasn't, which means I haven't moved. The greater the slope or gradient in this line, the greater the speed.

If we move from the average speed to the actual speed that we have at any given point, then we come up against the same problem as before. The two points through which we draw the line in the diagram get closer and closer to each other, as the distance covered that we are looking at gets smaller. Eventually, the two points come together and then it is no longer immediately clear how and with which gradient the line should be drawn. This line, which represents the gradient or in this case the momentary speed, is called the "tangent," and Descartes found a way to draw it. That was a remarkable achievement, but his solution was only practical for specific curves. The general problem of the mathematical explanation of changing dimensions was one he couldn't solve.

There were a few other mathematicians in the seventeenth century who made occasional small advances in the investigation of infinity or the drawing of tangents on curves. But these were isolated cases, which could only be applied to certain problems. What was missing was a new approach that would provide a comprehensive solution to the question and didn't only consider the calculation of speeds: always, when you want to investigate how a certain dimension changes when subject to the change of another dimension, you face the same problem. And such processes are to be found everywhere: the change in house prices subject to population growth; the change in the global temperature subject to CO_2 emissions; the change in the concentration of chemical elements during a molecular reaction; the change in the calculating time of a computer program subject to the processing power; the change in the mass of a rocket subject to its altitude. These and other phenomena can only be understood if the appropriate mathematical tools are available.

NEWTON'S DIFFERENTIAL AND INTEGRAL CALCULUS

It was precisely these tools that Isaac Newton created. But Gottfried Wilhelm Leibniz also created the same tools, and the question as to which of the two was the true originator of this powerful new mathematics, and whether one had copied from the other, occupied not only Newton and Leibniz for years, but also science historians (almost) up until the present day.

Newton's route to a mathematical understanding of change began—like so much else in his life—in the years 1665 and 1666. Having fled from the plague in Cambridge to his home village of Woolsthorpe, he didn't only think about optics and falling apples, but also about infinity. He struggled with this term, just as he had struggled to find terms for phenomena like "force" and "mass." For him, it wasn't only about pure mathematics, but also about the basis of matter itself: How far can one divide something? Is there anything that is indivisible? Can points with no spatial dimension still be put together in some way to form a line? What does "infinitely small" mean? Nothing—or actually something? And if something, then what?

Newton had to fight his way through a tangle of imprecise terms: "infinite," "uncertain," "indistinguishable." He used a special symbol (a small "0") in his texts to refer to this strange thing that was in a way nothing, yet was at the same time something after all. He wrote about lines that differ from one another by an infinitely small amount, and lines that do not differ from one another at all, and endeavored not to erase the almost indistinguishable difference between these descriptions.

The big breakthrough came when Newton no longer saw a curve as a rigid geometrical image, but rather as a point that moves and, so to speak, pulls the lines with it and draws them. He invented new words to refer to these flowing and flexible images: fluxions and fluents. Fluxions were Newton's way of being able to calculate a tangent after all, even if there was only a single point through which it could be drawn. Simplistically put, he just imagined a further point that was an infinitely small distance from the actual point. This distance, which didn't actually exist, was delineated by his fluxions, and from them he could then measure the gradient of the tangent. What Newton had created is learned by everybody in math classes today in the form of differential calculus. He also created the counterpart to this: what is known today as integral calculus, in which out of an infinite number of small areas, the area beneath a curve can be calculated.

Newton had thus found his tools. He was in a position to characterize changes mathematically and to avoid all the problems and paradoxes that had given people headaches since ancient Greek times. In one fell swoop, he was able to solve numerous mathematical problems which had previously been insoluble, and deal with questions much more quickly than had been possible to date. He used these tools to make the calculations that would form the basis of the *Principia*, but there was one thing that he didn't do: make his method public. The publication of these mathematical findings had been planned, but the criticism by Hooke and others that had followed his first publications put him off (see chapter 3). He did write everything up in 1671 in a paper titled *Tractatus de Methodis Serierum et Fluxionum*, but left this unpublished, just like his other mathematical book *De Analysi per Aequationes Numero Terminorum Infinitas*. That was at least, however, one of the texts he showed to his colleague John Collins, who was so enthused by it that, without Newton's knowledge, he made a copy of it (which would later lead to problems. . . .). Newton himself continued to use his method of fluxions, but removed every clue to their use from his work. He had achieved one of the most important breakthroughs in the field of mathematics—and remained silent about it.

LEIBNIZ, THE PLAGIARIST?

While Newton was thinking about infinity in Woolsthorpe, Gottfried Wilhelm Leibniz was just in the process of trying to get his doctorate. His dissertation was finished, but the University of Leipzig had still refused to confer his doctorate—

aged just twenty, he was still too young. That didn't mean that he hadn't achieved enough by then. Even as a child, he had been exceedingly thirsty for knowledge and had taught himself Latin and Greek at the tender age of eight using books from his parents' library. He began his studies at the University of Leipzig in 1661, when he was fourteen. Leibniz wanted to be a lawyer like his father, but also studied theology and philosophy. And mathematics, too: in 1663, he switched to the University of Jena for a semester, where he met Professor Erhard Weigel. Weigel was not really known as a famous mathematician, but he was an outstanding teacher. He was an advocate of clear and comprehensible language and encouraged his colleagues and students to avoid Latin arguments in specialist language and instead to use simple German. Weigel's influence was later to play an important role in Leibniz's bitter quarrel with Newton.[4]

In 1666, Leibniz would have been ready to receive his doctorate, but the university council in Leipzig was against this. So he went to Nuremberg, to the University of Altdorf (which no longer exists), and, after a few months there, received his doctorate,[5] "with great applause," as Leibniz himself reported. "In my public disputation, I expressed my thoughts so clearly and felicitously, that not only were the hearers astonished at this extraordinary and, especially in a jurist, unexpected degree of acuteness; but even my opponents publicly declared that they were extremely well satisfied."[6]

It would seem that the young Leibniz could not be accused of excessive modesty. But his work must indeed have met with great enthusiasm, since he was offered a professorship in Altdorf soon after graduating. He rejected this, however; he felt he was destined for a higher calling. He wanted to know more, learn more, and have contact with as many other like-minded thinkers as possible. In order to achieve this, he entered into the service of the Mainz Archbishop Johann Philipp von Schönborn as an advisor.[7] He then went to Paris in 1672 as a diplomat, where his job was to advise the French king Louis XIV (the "Sun King") and discourage him from going to war against Holland. Leibniz's alternative suggestion was a campaign against Egypt—an idea that found no favor with the king and was simply ignored. However, Leibniz did finally find suitable company in Paris, and he began to concern himself with a number of different subjects and to think seriously about mathematics.

Christiaan Huygens, who also lived in Paris at the time, recognized the potential in Leibniz. At a meeting, he gave him a problem to solve. Leibniz was asked to calculate the sum of the so-called "reciprocal triangular numbers,"[8] which he managed to do. Huygens then showed Leibniz which mathematical books to read and how he should further his education. In 1673, the young German traveled to England, where he met many members of the Royal Society. He presented to them one of his inventions: a mechanical calculating machine that could not only add and subtract like those already in existence, but also multiply and divide. In so doing, however, he also became the involuntary inventor of the demo effect—when he wanted to demonstrate his instrument, it didn't work. This didn't go down too well with the Royal Society (and Robert Hooke claimed, of course, that he could construct a much better machine). The episode with the calculating machine would later cause Leibniz more problems, as would a visit to the English mathematician John Pell's house. Leibniz apparently wanted to show off a bit there and performed a mathematical trick on the calculation of roots that he had developed himself. Pell then showed him a book in which the very same method had long since been published, a book that Leibniz could theoretically also have read. Leibniz claimed that he hadn't read it, but a slight doubt remained: had Leibniz here perhaps presented somebody else's work as his own? He quickly wrote an explanation, in which he denied any wrongdoing, and delivered it to the Royal Society. This was perhaps done with the best of intentions, but there was now a written document linking Leibniz to a potential case of plagiarism, something he would later regret.

For the time being, however, he thought no further of the matter, instead taking it as a reason to gain a judicious and comprehensive understanding of mathematics. This led him in the next few years to his big breakthrough: like Newton, he found a mathematical way to demonstrate changes. What

Newton called "fluxions," Leibniz referred to as "calculus"—
but it worked in exactly the same way, only perhaps a little
better. For Leibniz, shaped by his studies under Erhard Weigel,
gave plenty of thought from the beginning to how his new
mathematics could be described and used as simply as possible. Where Newton had been happy enough that the whole
thing worked, Leibniz also came up with a new and practical
language of symbols to make it as easy as possible for mathematicians to follow.

In 1676, Leibniz had more or less finished his work on his
"differential calculus" (and also, shortly afterward, his work
on "integral calculus"). Now a court counselor and librarian in
Hannover, he fine-tuned the details a little before publishing
his findings. In 1684, the first of two articles appeared in which
he presented the basic rules of his new method. Newton had
come up with differential calculus back in 1666, but the Royal
Society first learned about this new mathematics in 1685,
in a book by the Scottish mathematician John Craig, who
described Leibniz's findings in it.

This did not mean, however, that Leibniz and Newton had
not known of one another before then. They had indeed not
met personally, but had at least corresponded with each other.
To unravel and analyze the full interaction between Leibniz
and Newton, including all the information that was leaked to
them, or could have been, would go far beyond the limits of
this book (and generations of science historians have already

done that in every detail; see the bibliography). Newton wrote two letters directly to Leibniz in 1676, however, in which he praised his colleague in exceptionally polite terms. The two of them also corresponded about mathematics, though without revealing anything about their respective discoveries. It must have been clear to them both that the other had developed a similar mathematical method, but both of them assumed that it was just a similar method and not the same one.

At least on the surface, the two of them seemed to have no interest in quarreling about the original authorship. Leibniz considered himself the inventor of differential calculus, and the rest of the mathematical world seemed to as well. Newton knew that he had had the same ideas much earlier, but made no public utterances on the matter and continued to publish nothing. In the first edition of the *Principia*, he even specifically mentioned that he knew of Leibniz's method.

NEWTON BRINGS OUT
THE HEAVY ARTILLERY

The now-famous row about the original authorship of the new mathematics was actually instigated from the second rank of scientists. Nicolas Fatio de Duillier, formerly Newton's close friend, published a letter in 1699 in which he directed at Leibniz a barely veiled accusation of plagiarizing

Newton's work. He said that Leibniz had heard of Newton's work during his time in England and only then developed the idea for his own method. Leibniz immediately published a response in *Acta Eruditorum*, a journal that he himself had founded (and the first German scientific journal), in which he forcibly denied the accusation. He then also wrote a kind of review of his own response, which he also published in the journal, though this time anonymously. In addition, he sent a letter of complaint to the Royal Society, of which he was now also a member. Newton kept out of the whole matter to begin with. This first attack on Leibniz was therefore without success, but the battle was by no means over.

In 1703, Isaac Newton was elected president of the Royal Society. He now seemed to feel more secure and began publishing works again. In 1704, *Opticks* appeared, a work that was largely concerned with his optical research, but which also contained an appendix at the end in which he publicly presented his mathematical method of fluxions for the first time. Leibniz reviewed this appendix and published his opinion, again anonymously. In his text, he mostly praised himself and hinted that Newton might have copied from him.

The row escalated in 1708, when the physicist and mathematician John Keill published a text in which he directly accused Leibniz of plagiarism. Leibniz didn't really know what to think of this. From his point of view, Keill wasn't a worthy opponent, and was merely a mediocre mathematician, and Leibniz didn't wish to waste time on him. He therefore did what he had done before in response to Fatio and addressed himself to the Royal Society, demanding that the "upstart" (as Leibniz wrote to the Society's secretary) Keill should publicly apologize for this defamation. This time, however, he had misjudged the situation, for now Newton was president and had in the meantime heard of Leibniz's anonymous review of his work, complete with its hint at plagiarism. For Newton, the time for holding back had passed, but for the moment, he left it to Keill to speak for him.

Keill subsequently repeated his accusations, rather than apologizing, and in 1712, the Royal Society finally set up a commission to clear up the matter. It was not a particularly fair business from the beginning. Newton was still president of the Royal Society, and Leibniz didn't get a chance to tell his side of the story to the commission. All the old stories were rolled out once more: how Leibniz had so boastfully appeared with his calculating machine, which then didn't work after all; how he had showed off to John Pell with mathematical tricks that others had discovered before him; how he had had the opportunity during his visit to England to look at private documents with Newton's work on mathematics. Hardly surprisingly, the commission soon came to the conclusion that Newton had invented the new mathematics, that he had been the first. In the concluding report, it was again put on record that Keill had said nothing wrong and that Leibniz had certainly had the opportunity to plagiarize Newton.

Now Leibniz was truly angered. He and his friend, the Swiss mathematician Johann Bernoulli, thenceforth insulted the English in their letters, calling them "men full of vanity, who have always used every opportunity to present German insights as their own" and saying they were "envious of all other nations." In 1713, Leibniz then published a pamphlet—again anonymously—that became known under the name "Charta Volans." In it, he praised himself and spoke of his "honest nature," which he took as a standard by which to judge others. That is why Leibniz, wrote the anonymous author (i.e., Leibniz himself), had never thought that Newton might have copied him. The reason for Keill's attack was the "unnatural hostility towards foreigners of the English" and Newton allowed himself to be influenced by sycophants who had no idea of the true course of events. In addition, his desire for fame was a "sign of a mind that was neither respectable nor honest." As evidence of all this, Leibniz mentioned the quarrels Newton had had with Hooke and Flamsteed.[9]

Now it was Newton's turn to get into a rage. He himself wrote a comment on the Royal Society commission's report, also anonymously, in which he attacked Leibniz once more, this time not only declaring his precedence with regards to differential calculus, but also discrediting Leibniz's special symbol language.

"Mr. Newton," wrote Newton, talking about himself in the third person, "doth not place his Method in Forms of Symbols, nor confine himself to any particular Sort of Symbols for Fluents and Fluxions."[10] Then he showed precisely the confused notation that characterized his work: "And where he puts the Letters x, y, z for Fluents, he denotes their Fluxions, either by other Letters as p, q, r; or by the same Letters in other Forms as X, Y, Z or $\dot{x}$, $\dot{y}$, $\dot{z}$; or by any Lines as DE, FG, HI," before proudly asserting: "All Mr. Newton's Symbols are the oldest in their several Kinds by many Years." In the last point, Newton may even have been right. His predilection for geometry had already manifested itself in the *Principia*, where he had presented the results gained through infinitesimal calculus not as such, but rather translated back into the old mathematical language of geometry. The rest, however, was merely his own personal opinion, and his notation (for him perhaps clear, but for the rest of the world rather confusing) did not prove to be useful in the long run. Formulating the new mathematics which he and Leibniz had developed with the old symbols and conventions made the whole thing too complicated for everyone who wasn't a genius. The new symbols that Leibniz invented,[11] and the clear language formulated from them, therefore played a major role in the breakthrough of calculus and in helping other mathematicians to accept it.

The mud-slinging progressed to the next round: Leibniz wrote to a friend that he would love to attack Keill physically, instead of just with words. In public, however, he resorted to criticizing Newton's physics. The theory of universal gravity meant nothing to Leibniz. He considered the idea of forces that exerted an effect over great distances across empty space to be nonsense and wrote to Bernoulli: "I have tested it and had to laugh at the idea. . . . This man is not very successful in metaphysics." Leibniz considered empty space to be impossible and was an advocate of Descartes' vortex theory. The conflict about physics and the nature of space was an attempt to shift the row with Newton from mathematics to philosophy. Here, he considered himself—quite rightly—to be superior to Newton and hoped therefore to gain an advantage.[12] In the end, though, he merely did himself damage with this tactic.

Leibniz died in Hannover on November 14, 1716. Even after the death of his opponent, Newton was undeterred and continued to publish new attacks. In the second and third editions of the *Principia*, he deleted all mention of Leibniz. His fame increased ever more, and he was feted even during his lifetime as one of the greatest geniuses of all time; against him, the dead Leibniz had no chance. His criticism of Newton's universal gravity was one of the few cases where he was completely off the mark. And the more successful Newton's new physics became, the more people came to accept the opinion that Leibniz might have been wrong with mathematics too, if he had misunderstood physics. Conspiracy theories arose: hadn't Henry Oldenburg been secretary of the Royal Society at the time when Leibniz was visiting London? And wasn't Oldenburg a German like Leibniz? Perhaps he had been a spy for his country and secretly passed on Newton's findings. . . .

Newton seemed to have won the battle for differential calculus. But over the centuries, a more differentiated (how apt!!) image developed. It took a long time for the conflict to finally end. After Newton's death, there were plenty of others among the following generations of scientists who took his side and continued to fight for his fame and recognition. British scholars in particular refused to accept a bad word about their "national saint," and it was only slowly that people came around to the notion that Newton wasn't the godlike genius who had been portrayed during his lifetime and, above all, in the decades after his death. Slowly but surely, historians began to reveal not only Newton's complicated and unpleasant character, but also the false accusations and unfair attacks on Leibniz: "The great fault, or rather misfortune, of Newton's life was one of temperament; a morbid fear of opposition from others ruled his whole life . . . when he became king of the world of science it made him desire to be an absolute monarch; and never did monarch find more obsequious subjects. His treatment of Leibniz, of Flamsteed . . . is, in each case, a stain upon his memory," writes the English mathematician Augustus De Morgan in a biography of Newton from the year 1846.

Today, the long confrontation is merely an interesting episode in science history, and there is—with the exception of a few details—no more dispute about the true course of events. The vast majority of science historians take the view that Newton and Leibniz made their mathematical discoveries independently of each other. Newton was the first to do so, but Leibniz was the first to actually publish his findings. In any case, his language of symbols was superior to the complicated notation of Newton's fluxions, which is why it is still used today.

Imagine for a moment a world in which two such great minds as Newton and Leibniz were friends who supported and encouraged each other, instead of quarreling and holding each other back: in such a world, Newton wouldn't have kept his work hidden from the world and, together with Leibniz, could have taken mathematics to unparalleled heights. Instead, though, two of the greatest scientists of all time had nothing better to do than to dish the dirt on each other.

THE CRUX OF HAVING TO BE THE FIRST

This time, things are quite clear. Pretty much everything that Isaac Newton did in this particular episode definitely shouldn't serve as an example for modern science. The same is true for Leibniz. The problems that the two of them had to face still exist, of course. The quarrel between Newton and Leibniz revolved mainly around precedence, which is a concern that is perhaps even more important today than back then. Whether somebody has been the first to discover or find out something or not is a determining factor in scientific careers and important prizes and awards. The mistake that Newton made—keeping his research results to himself for years, or perhaps even decades—is one that people today can hardly afford to make.

The world of science is now much more interlinked than in the seventeenth century. So many people are working on big projects that hardly anything can be kept secret and competitors would find out much more quickly than in Newton's day. When, for instance, the scientists at the LIGO Gravitational-Wave Observatory, after decades of trying, finally provided evidence of gravitational waves in autumn 2015, it only took a few weeks for the first rumors to spread on the internet. And when the discovery was officially reported in February 2016, everyone already knew beforehand what was going to be announced. In this case, it wasn't so serious, since nobody else would have had the technical wherewithal anyway to profit from the data and thus get the drop on LIGO.

But when it comes to the search for asteroids or comets, for example, you often find today the same secretive behavior as in the past. In December 2004, a team of astronomers under the American Mike Brown discovered a large asteroid in the outer solar system. The object had a diameter of almost 1,000 km and is classified today as a dwarf planet with the name "Haumea." Brown and his team didn't make their discovery public right away, however. They waited until July 2005 and even then only made an announcement that pointed to a future publication. In this preview for the discovery, they gave no concrete data about Haumea, but used a codename for the asteroid that they had invented themselves. What Brown and his colleagues had overlooked was that, if you did an internet search for this codename, you got the logbooks of the telescopic observations—which had actually been intended for internal use only. This enabled people to find out Haumea's exact position in the sky. Shortly after Brown's announcement, Spanish astronomers led by José Luis Ortiz Moreno suddenly went public and declared that they had discovered a large celestial object in the outer solar system.

The whole situation proceeded a bit like the row between Newton and Leibniz. Brown wrote to Ortiz and wanted to know if he had stolen his data. Ortiz didn't answer, so Brown called the International Astronomical Union (IAU) to ask them to investigate the matter. Examination of internet records revealed that Brown's data had indeed been accessed by a computer in the region of the Spanish observatory on the day before the Spaniards' public announcement. The Spanish then admitted that they had accessed the logbooks, but said they had only done so to verify their own observations; they had already discovered the asteroid themselves before and merely wanted to check what their colleagues had found. Brown doubted this, but it has never become completely clear what actually happened. The IAU gives the date of the Spanish team's announcement as the official discovery date. The right to name the celestial body, however, was granted to Mike Brown, although this is normally the exclusive preserve of the discoverers.

Similar disputes about precedence are constantly taking place in the world of science. They can be about major, revolutionary results[13] or mere bagatelles that don't attract much public attention. Unlike Isaac Newton, however, everyone is fully aware that it is a dangerous business not to make your data public. The longer you wait to do so, the greater the risk that you will be overtaken by others.[14]

The desire to produce results as quickly as possible can lead to plagiarism in extreme cases. Both Newton and Leibniz certainly didn't behave in an exemplary fashion, but there has so far been no concrete evidence that either of them did indeed copy from the other. Presenting other people's insights as your own is just as reprehensible and unethical in the world of science (and not only there!) today as it was in the seventeenth century. Nevertheless, plagiarists are constantly being found guilty—very often in cases where those involved are not so much interested in an academic career, but rather in an academic degree. When the German defense minister Karl-Theodor zu Gutenberg was forced to give up his doctor's degree in February 2011, because it was proven that large sections of his doctoral thesis were not his own work and had instead been taken from other authors without acknowledgment, this led to a whole wave of checks of academic papers by politicians in Germany. As a result, the ministers of the European Union Silvana Koch-Mehrin and Jorgo Chatzimarkakis had to give up their doctoral degrees, as did the Christian

Democratic Union politician Matthias Pröfrock and (ironically enough) the minister for education and research Annette Schavan. They certainly won't be the only politicians to have been less precise with their work than they should have been, and there are bound to be cases of plagiarism in other sections of the population, too. For such cases, at least, it would be very simple to prevent future attempts at plagiarism: it would simply be necessary to get rid of the use of the title of "doctor."

The politicians named above were obviously not interested in an academic career and it can be presumed that they only wrote their dissertations because of the prestige attached to the title. For scientists, on the other hand, it doesn't really matter whether they can put a "Dr." in front of their names or not. What matters for them is their actual work, the results that they come up with. If you read the papers in the scientific journals carefully, you'll realize that the authors only appear under their names and academic titles are not to be seen. Whether a scientific piece of work is good or bad is determined by testing it against reality, not by the (supposed) authority of a title. It would therefore be absolutely no problem for the scientific community if the title of doctor was simply abolished. It isn't required to prove your education; after all, you get certificates and statements of attendance for all the lectures you attend, and the scientific value of your thesis doesn't change if it is no longer called a "doctoral thesis." The only thing that would disappear with the abolition of the doctor's title would be the motivation to gain supposed authority with illegal plagiarism.

It will take a long time until the rather conservative world of the universities decides to take such radical steps, however. Until then, it is all the more important not to yield to the temptation of passing off other people's work as your own. And perhaps Newton and Leibniz can serve as an example here. The two of them used a whole load of dirty tricks in their row about the precedence of differential and integral calculus. Presumably, however, neither of them would have dreamt of replacing true research with ignominious copying. Both Isaac Newton and Gottfried Wilhelm Leibniz had too great a thirst for knowledge to do that. Of course they were interested in recognition and fame. Naturally they were angered by the prospect of somebody else getting the praise for their own achievements. And of course they were both keen to be viewed favorably by posterity. Newton and Leibniz may have been geniuses, but they were also human. Both of them spent almost their entire lives deciphering the mysteries of nature, consistently and with little regard for losses to themselves (or others). Both of them wanted knowledge at all costs— knowledge about everything there is to know. Neither Isaac Newton nor Gottfried Wilhelm Leibniz would have dreamt of copying something from somewhere and thus denying themselves the pleasure of finding it out for themselves. They may well have been assholes. But they were also genuine scientists.

CONCLUSION

It may be a slight exaggeration to claim that Isaac Newton alone laid the foundations for modern physics. But only a slight one. His work is full of innovative ideas that, to begin with, were as incomprehensible for his contemporaries as they are indispensable for scientists today. Take the idea of a force that acts across empty space. Newton formulated the force of gravity as just such a force, and his contemporaries had great difficulty with this; they couldn't imagine that a force could act "just like that," without there being any material contact between the objects involved. It seemed like magic, and even Newton wasn't completely sure whether some divine intervention was perhaps necessary after all to convey the force of gravity.

As in so many cases, this idea of Newton's proved to be a fruitful starting point for further research. Modern physics has solved the problem of forces acting "just like that" by showing that there are fields of force. An object can create a field around itself that spreads out and influences other objects. The force of gravity can be exerted in this way, as can other forces like electricity or magnetism. The mastery and manipulation of electrical and magnetic fields form the basis of our modern technology, but even the fundamental characteristics of matter itself are explained by fields in physics today. Quantum field theory defines particles as the agitation of fields: photons, i.e., light particles, arise from the agitation of electromagnetic fields. Other particles are created out of their own particle fields; the so-called standard model of particle physics explains the interaction between all of these fields.

Isaac Newton also took the first step toward another major project of modern science: the unification of all forces acting in nature. His law of universal gravitation showed that completely different phenomena in the sky and on the earth can be explained by a single force. If we just observe the world in the right way, the differences disappear and everything appears much simpler and more elegant than before. In the nineteenth century, the Scottish physicist James Clerk Maxwell demonstrated that electricity and magnetism are also not two isolated phenomena, but are merely two different ways in which the unified force of electromagnetism can be exerted. Albert Einstein spent the last decades of his scientific career trying to find a way to bring electromagnetism and gravity together. Here, too, in a sense, he was following Newton's directions. In the book *Opticks*, Newton wonders: "Are not gross Bodies and Light convertible into one another, and may not Bodies receive much of their Activity from the Particles of Light which enter their Composition?" That almost sounds a bit like Einstein's most famous formula ($E = mc^2$), which demonstrates how matter and energy can be converted into one another. Einstein didn't manage to achieve a complete unification of the forces of mass and light, however, and the researchers who have followed him have also not yet succeeded in merging all forces known in nature into a single theory. Centuries ago, Isaac Newton took the first step toward a universal unification in physics, and every new generation since then has taken a small step further along this path. How long the path will be cannot yet be said; the search for the explanation of the universe that the media likes to call the "theory of everything" is not yet over.

In the seventeenth century, Isaac Newton provided the world with a research program on which the whole of science is still working just as intensively today as he did back then on his own. He was an eccentric, an egoist, a troublemaker, and a mystery-monger. He tolerated no criticism and was uncompromising, vengeful, and conniving. But he was also the greatest genius ever to have lived. No other scientist has had such an important, wide-ranging, and enduring influence on the entire world as he did. Sometimes, it would seem, if you want to change the world, you have to be both a genius and an asshole.

ACKNOWLEDGMENTS

I'm not sure whether Isaac Newton would thank me for writing a book about him that, while it does place great emphasis on his life and work, places equal emphasis on his undesirable character traits and even calls him an "asshole." He most probably wouldn't—I, however, am only too pleased to express my thanks to him, whom I consider to be one of the greatest geniuses of all time and whose work forms the basis of my own work and career as an astronomer. Without Newton, I couldn't have become an astronomer and couldn't have written this book, so thanks a lot, Isaac Newton!

I'd also like to thank my editor, Christian Koth, who was of great assistance in ensuring that I didn't get lost in the vast quantity of material and stories about Isaac Newton, and without whom this book would not have ended up as it is.

I am also grateful to all of my colleagues at the Science Busters; over the course of many discussions and appearances, they have helped me to understand how to present complex specialist topics as entertainingly as possible.

Thanks to Thomas Posch from the Vienna University Observatory for the wonderful guided tour of the museum there, for answering my questions on the history of astronomy, and giving me the chance to browse through books that Kepler once browsed through.

And finally, I'd like to thank the test readers, who reliably informed me when things were not clear enough: many thanks to Dagmar Fuchs, Nina Wallerstorfer, André Lampe, Florian Rodler, Matthias Kittel, and Franziska Hufsky.

Thank you!

BOOKS AND SOURCES

Few other scientists have been the subject of as many books as Isaac Newton, and there is also a plethora of works about his contemporaries and the development of science in the seventeenth century. There is no scope here for a complete overview of all the literature concerning Newton, but I would like to mention a few important works that I used myself when researching this book.

The definitive biography is still *Never at Rest* by Richard Westfall. If the nine hundred–odd pages are a bit too much, I can also recommend the 2003 book *Isaac Newton* by James Gleick as an introduction to Newton's life. Gleick's book is short and readable, yet still includes all the most important aspects of Newton's life and offers an extensive list of source material. *Newton's Gift: How Sir Isaac Newton Unlocked the System of the World* by David Berlinski is also short, though written in slightly more mathematical and scientific language, and describes Newton's main achievements in science. Another biography, which above all deals in detail with Newton's alchemistic work, is *Isaac Newton: The Last Sorcerer* by Michael White. *The Foundations of Newton's Alchemy* by Betty Jo Teeter Dobbs deals exclusively with Newton's alchemy and the history of alchemy itself, but its complexity means that it can only really be recommended for readers interested in this particular subject.

Those who really wish to immerse themselves in Newton's science can of course read his actual publications and letters. At http://www.newtonproject.sussex.ac.uk, the original and translated versions of Newton's works have been made public. However, Newton's magnum opus—his *Principia*—is presented in a much more comprehensible form in Colin Pask's *Magnificent Principia*. Don't let a bit of mathematics and corresponding diagrams and formulae put you off—you'll be rewarded with a generally understandable explanation of the most important aspects of Newton's work.

The fascinating story of Newton's estate is told by Sarah Dry in her book *The Newton Papers: The Strange and True Odyssey of Isaac Newton's Manuscripts*. The row between Isaac Newton and John Flamsteed (which is of course also described in all the above-mentioned biographies) is vividly recounted in the short book *Newton's Tyranny* by David and Stephen Clark. The story of a supposed dispute between Isaac Newton and the physicist Stephen Gray, which is also told in this book, seems to be the product of the authors' imagination, however, and does not come from the available sources.

Thomas Levenson writes about Newton's career in finance and his pursuit of counterfeiters in his highly readable book *Newton and the Counterfeiter*, which also contains an extensive biography of William Chaloner. The short book *The Calculus Wars* by Jason Socrates Bardi is also well worth reading; it deals with the row between Isaac Newton and Gottfried Wilhelm Leibniz. You can find out all about Robert Hooke's life in both Stephen Inwood's *The Man Who Knew Too Much* and Lisa Jardine's *The Curious Life of Robert Hooke*.

The collection of essays called *Seeing Further: The Story of Science, Discovery, and the Genius of the Royal Society*, edited by Bill Bryson, provides an overview of the history of the Royal Society, while I would particularly recommend *The Clockwork Universe: Isaac Newton, the Royal Society, and the Birth of the Modern World* by Edward Dolnick, which gives a fascinating, informative, and easily understandable overview of the development of natural sciences in Newton's day.

IMPORTANT DATES IN ISAAC NEWTON'S LIFE

Unless specified otherwise, the dates refer to the Gregorian calendar in use today.

December 1642: Birth of Isaac Newton (according to the Julian calendar used at the time).

January 1643: Birth of Isaac Newton (according to the Gregorian calendar used today).

January 1646: Newton's mother Hannah marries Barnabas Smith (Newton's biological father having died three months before the birth of his son).

1655: Newton attends school in Grantham and boards with the local apothecary.

1659: At his mother's request, Newton has to return to Woolsthorpe.

1661: Newton begins his studies at Trinity College, Cambridge.

1665: Newton completes his studies in January; in August,

the outbreak of the plague forces him to return to Woolsthorpe, where he remains for eighteen months. Here, he investigates optics, mathematics, and gravity, and lays the foundations for his later, revolutionary works.

1667: Newton is elected a fellow of Trinity College.

1668: Newton visits London for the first time.

February 1669: Newton describes his reflecting telescope in a letter to the Royal Society.

October 1669: Newton becomes Professor of Mathematics at Cambridge.

January 1672: Newton is elected a member of the Royal Society.

February 1672: Newton sends his findings about the nature of light to the Royal Society. His row with Robert Hooke begins.

February 1675: Newton attends a meeting of the Royal Society for the first time.

1679: Newton and Hooke begin correspondence about the motion of celestial bodies.

1680: Newton observes a comet.

1681: Newton begins correspondence with John Flamsteed about the motion of comets.

January 1684: Halley, Hooke, and Wren meet in a coffee house and discuss a law to describe the motion of the celestial bodies.

August 1684: Edmond Halley visits Newton in Cambridge

and prompts the latter's work on gravity. Newton begins work on the *Principia*.

April 1686: Newton presents the first volume of the *Principia* to the Royal Society.

July 1687: The *Principia* is published.

July 1693: Newton suffers a nervous breakdown.

September 1694: Newton visits John Flamsteed in Greenwich. The row about the star catalogue begins.

1696: Newton is offered a post at the Royal Mint.

February 1700: Newton becomes Master of the Mint.

November 1701: Newton is elected to Parliament.

December 1701: Newton resigns from his post as Professor of Mathematics at Cambridge.

November 1703: Newton becomes President of the Royal Society.

1704: Newton publishes *Opticks*, with his work on calculus first appearing there in an appendix.

1705: The precedence row with Leibniz begins.

April 1705: Newton is knighted by Queen Anne.

March 1712: The Royal Society sets up a committee to clear up the precedence row between Newton and Leibniz.

1713: Newton publishes the second edition of the *Principia*.

1726: Newton publishes the third edition of the *Principia*.

March 1727: Isaac Newton dies.

NOTES

Introduction

    1. From 1582, most countries in Europe used the Gregorian calendar that was introduced that year (and is still used today). In England, however, people could only bring themselves to use it in 1752. (Unless otherwise stated, the dates in this book refer to the Gregorian calendar used today, such as the date of Newton's birth, January 4, 1643.)

    2. I therefore deliberately decided to forgo a chronological presentation of Newton's life, and instead jump back and forth in time, in order to look at what I consider to be the interesting episodes in his life and work from the specific angle of this book.

    3. See "Newton's Notebooks," Newton Project, http://www .newtonproject.ox.ac.uk/texts/notebooks.

    4. Unless you're a religious fundamentalist.

Chapter 1. At All Costs: Newton, Ruthless in the Extreme

    1. Letter from Humphrey Newton to John Conduitt, January 17, 1727/28. See "Two Letters from Humphrey Newton to John Conduitt," Newton Project, http://www.newtonproject.ox.ac.uk/view/texts/ normalized/THEM00033.

    2. Ibid.

3. See "Newton's Notebooks," Newton Project, http://www
.newtonproject.ox.ac.uk/texts/notebooks.

4. See "An Extract of a Letter Not Long since Written from Rome,
Rectifying the Relation of Salamanders Living in Fire," *Philosophical
Transactions* 1 (1667): 377–78, http://rstl.royalsocietypublishing.org/
content/1/21/377.1.short. The seventeenth century was certainly not a
time of brief, concise titles. Today, the title of the text would probably be
something like "He threw the salamander into the fire—you won't believe
what happened next!"

5. "An Extract of M. Dela Quintiny's Letter, Written to the
Publisher in French Sometime Agoe, Concerning His Way of Ordering
Melons; Now Communicated in English for the Satisfaction of Several
Curious Melonists in England," *Philosophical Transactions* 4(1669): 901.

6. The pressure of an ideal gas is inversely proportional to the
volume, if the temperature remains the same.

7. The French words *thé*, *caffé*, and *chocolat* had just begun to be used
in the German language at the time; the dog wasn't bilingual, therefore,
merely very well-informed about the latest culinary developments.

8. But honestly—please don't look directly at the sun! It really is
dangerous, even using sunglasses, thermal blankets, colored bits of glass, or
other supposedly safe methods. Get hold of a pair of eclipse glasses, ideally
NOW and not merely a day before the next major astronomical event. Or
observe the eclipse in an observatory: the people there will know their stuff
and will have prepared appropriate means of seeing the eclipse. You should
only look directly at the sun if you have already seen everything else there is
to see and have no need to see any more. Or if you are Isaac Newton.

9. Although, in view of his extensive alchemistic experiments (see
chapter 6), it would certainly be tempting to invent a suitable fictional
biography, in which a failed experiment turns the puritanical scientist into
a musclebound comic-book superhero.

10. It is likely that Newton had health problems at the time. Whether

it was full-blown depression, overwork, or something else cannot be said for sure today. But he was certainly in the mood for a change of circumstances.

11. Newton's contemporary, the French philosopher Voltaire, remarked, "Fluxions and gravitation would have been of no use without a pretty niece."

12. A groat was a small silver coin worth four pence.

13. *Guzman Redivivus. A Short View of the Life of Will. Chaloner, the Notorious Coyner, Who Was Executed at Tyburn on Wednesday the 22d of March, 1698/9* (London: J. Hayns, 1699).

14. See Thomas Levenson, *Newton and the Counterfeiter: The Unknown Detective Career of the World's Greatest Scientist* (London: Faber and Faber, 2009).

15. See Levenson, *Newton and the Counterfeiter*, and "William Chaloner's Letter to Isaac Newton," Newton Project, http://www.newtonproject.ox.ac.uk/view/texts/normalized/MINT00918.

16. This is what we call the search for external sources of funding, e.g., from foundations, research funding organizations, or business enterprises.

Chapter 2. *Principia* First: Newton the Egoist

1. See David Clark and Stephen Clark, *Newton's Tyranny* (New York: W. H. Freeman, 2000).

2. The Royal Observatory in Greenwich, which still exists today, though it is no longer used as a scientific observatory.

3. The measurements were of course still relatively imprecise. If you really want to know exactly where you are, you need to take many different factors into account about which much too little was known back then: that the earth is not a perfect globe, for example, and actually has an irregular form, or that the earth's axis sways to and fro a little thanks to the influence of the moon.

4. Although, to be precise, that actually depends on what the rotation is in relation to. In relation to the starry sky (if we assume that this remains unchanged), the earth only takes 23 hours, 56 minutes, and 4.099 seconds for one full rotation.

5. See *Newton and Flamsteed. Remarks on an Article in Number CIX of the Quarterly Review* (Cambridge and London: 1836), and Clark and Clark, *Newton's Tyranny*.

6. See Clark and Clark, *Newton's Tyranny*.

7. Richard S. Westfall, *Never at Rest: A Biography of Isaac Newton* (Cambridge: Cambridge University Press, 1980).

8. Richard S. Westfall, *The Life of Isaac Newton* (Cambridge: Cambridge University Press, 1993).

9. And Flamsteed received a second humiliation: Newton's friend, the astronomer Edmond Halley, received a fee of 150 pounds for the processing of Flamsteed's data.

10. Richard S. Westfall, *The Life of Isaac Newton* (Cambridge: Cambridge University Press, 2015).

11. Flamsteed later makes explicit reference to being called a "puppy" by Newton. What Newton's aim was in calling him this is not clear, but that's probably what really bad insults were like in those circles back then.

12. Stealing the data from its originators like Newton did, however, is something to be avoided.

Chapter 3. False Modesty and Easily Offended: Newton the Shrinking Violet

1. Astrology was of course just as much nonsense then as it is now. But if you wanted to practice astrology, you had to work everything out for yourself, since the computer programs that allow anybody to put together their own horoscope today didn't exist at the time. In order to understand

astrology in Newton's day, therefore, you needed to have a command of mathematics.

2. One example is the so-called geometric series: $1 + \frac{1}{2} + \frac{1}{4} + \frac{1}{8}$ + etc. Although an infinite number of figures is added here, the result is simply 2, since the figures get smaller and smaller.

3. René Descartes, *Principles of Philosophy*, XXVI.

4. Isaac Newton to John Collins, Trinity College, Cambridge, January 19, 1669, in *Correspondence of Scientific Men of the Seventeenth Century, Including Letters of Barrow, Flamsteed, Wallis, and Newton, Printed from the Originals in the Collection of the Earl of Macclesfield* (Oxford: Oxford University Press, 1841), p. 286.

5. And a ray of light that was "another green" would probably be just as much "another green" as before.

6. Such mirrors have little to do with the glass mirrors we know from our bathrooms, however. They were made of metal that was covered with a reflective layer.

7. The magnification is not, however, such an important feature of a telescope – at least not in modern astronomy. The stars and galaxies are much too far away for us to get a significantly larger image of them, regardless how strong the magnification is. The important thing in astronomers' telescopes is for them to collect more light than the human eye can, in order to be able to detect objects shining more and more weakly.

8. Isaac Newton to Henry Oldenburg, Cambridge, January 6, 1671, in *Correspondence of Scientific Men*, p. 311.

9. It was also Hooke who gave the word "cell" its biological meaning. When observing bits of plants, he noticed the small, separate sections, which reminded him of the narrow rooms occupied by monks in a monastery, and so called them "cells."

10. Probably "green," "another green," and "a light green."

11. What modern color theory calls "complementary colors."

12. Incidentally, the reviewers are not paid for their work. The

authors also don't get any money for publishing their results in a journal (indeed, they themselves often have to pay the publishers for this). That is a slightly absurd state of affairs and creates a whole new set of problems.

13. Thanks to my colleague's criticism, I did have to rewrite large chunks of my work and redo many things. Finally, however, I was able to produce a much better article than before, and also got it published ("Planets of Beta Pictoris revisited," *Astronomy & Astrophysics* 466 [2007]—in case anybody is interested).

14. Albert Einstein, in a letter to *Physical Review*, July 27, 1936, in Daniel Kennefick, "Einstein Versus the *Physical Review*," *Physics Today* 58, no. 9 (September 2005): 43.

15. Both Einstein and the reviewer (the American mathematician and physicist Howard Robertson) were proved right in February 2016, when the LIGO collaboration announced the first direct evidence of gravitational waves.

Chapter 4. Gravity without the Apple: Newton's Pugnacious Side

1. The buoyant force acting on an object in a medium is equal to the weight of the medium displaced by the object.

2. Isaac Newton, *Questiones quædam Philosophiæ*, in Newton Project, http://www.newtonproject.ox.ac.uk/view/texts/normalized/THEM00092.

3. For Newton, on the other hand, it was precisely the "second" that was an unusual term. Measurements accurate to the second were remarkable then, since there were scarcely any clocks that could measure time so accurately. Galileo himself used a pendulum to measure time.

4. Wilhelm Foerster, *Die Erforschung des Weltalls*, in Hans Kraemer, *Weltall und Menschheit* (Berlin and Leipzig: Band III, Verlag Bong and Company, 1903).

5. A comet would later be named after him: the famous Halley's Comet, which will next appear and be visible on Earth in 2061. Unless the world comes to an end before then, of course—see chapter 6.

6. It has to travel a greater distance in the same time. A point on the earth's surface describes a circle in one day that is exactly equal to the distance around the earth. The top of the tower is higher, and so the circle it describes has a larger diameter.

7. An astronomer and founding member of the Royal Society, who became famous above all as an architect. His most famous building is St. Paul's Cathedral in London.

8. A then unpublished work of Willughby's still in existence today concerns games in the seventeenth century. Here, there is the first use of the word "goal." According to Willughby, football is played on "a close that has a gate at either end. The gates are called goals." He also gave thought to the rules, since the players "often break one another's shins when two meet and strike both together against the ball, and therefore there is a law that they must not strike higher than the ball."

9. *De historia piscium libri quatuor* (1686).

10. Today, however, Willughby's book is a highly-prized collector's item, and each copy is worth several thousand dollars.

11. And later a further twenty copies, as a kind of bonus. You can just imagine how Halley's wife (to whom he had been married for four years at the time) reacted when her husband brought home another pile of books about fish instead of money.

12. At least back then.

13. This is the logical fallacy of the so-called Galileo gambit, which asserts that, since Galileo Galilei was ridiculed and criticized by everyone because he claimed that the earth revolved around the sun, although he was a genius and was absolutely right, then if I am ridiculed and criticized by everyone, then I must also be right.

Chapter 5. The Silent Revolutionary: Newton the Mystery-Monger

1. See Stephen Hawking and Werner Israel, eds., *300 Years of Gravitation* (Cambridge: Cambridge University Press, 1996).

2. See Colin Pask, *Magnificent Principia: Exploring Isaac Newton's Masterpiece* (Amherst, NY: Prometheus Books, 2013), pp. 124–26.

3. Although that is only a short version, to be precise. Newton's formulation is more general: force is the temporal change of momentum. Momentum is the product of mass and velocity and, therefore, changes in both velocity and mass must be taken into account in its temporal change. In the vast majority of cases, the mass of an object doesn't change, and then the well-known formula "Force is mass times acceleration (i.e., change in velocity)" does hold true. But sometimes—for example with rockets being launched, which burn large quantities of fuel within a short time and thus lose mass—Newton's more general formulation must be used.

4. In addition to these three axioms, Newton also came up with what is known today as the superposition theory, which explains how several forces acting on an object can be put together to make one single force.

5. Not for nothing is weightlessness in space also referred to as "free fall": an astronaut is not at rest, but rather falls constantly around the earth, like the moon.

6. A unit of length used by Isaac Newton that corresponds to approximately 32.48 cm. There was no uniform system of units at the time.

7. In mathematics, the branch of mathematics developed by Newton is also called "mathematical analysis."

8. The Mediterranean is not large enough for this, and the form of the coastline, unlike in the North Sea, for example, does not allow the water to mount up very much at high tide.

9. A precise explanation of the tides is extremely complicated and

Newton too was not in a position to explain all the details. This was only done by the physicists and mathematicians who followed him. Today, while we have managed to understand the tides, it is still incredibly difficult to give a clear explanation. Even in many textbooks, you find explanations that are wrong or so simplified that they give a distorted impression of the facts.

10. Not to be confused with the word "precision," which means something completely different. "Precession" is a specialist term that originates in the Latin word for "advance."

11. Contrary to widespread belief, there was never a notable number of scholars who thought the earth was a disk. This is a myth that developed relatively late. Even in the (supposed) "dark" Middle Ages, people knew about the earth's spherical shape.

12. The values as measured today are 6,378.137 km to the equator and 6,356.752 km to the poles.

13. Here, too, we can assume that Newton fiddled with the figures a little. With all the simplifications of his models and the quality of the observation data, it would be a huge coincidence if the theoretical calculations and observations tallied so precisely.

Chapter 6. In Search of the Philosopher's Stone: Newton's Esoteric Side

1. The full quote from a lecture by Keynes in 1942: "Newton was not the first of the age of reason. He was the last of the magicians, the last of the Babylonians and Sumerians, the last great mind that looked out on the visible and intellectual world with the same eyes as those who began to build our intellectual inheritance rather less than 10,000 years ago."

2. Newton didn't have to resign after all; the king passed an exemption that allowed him to remain at the university even without joining the clergy.

3. Yahuda MS 7.3, Jewish National and University Library, Jerusalem.

4. Although "Jehova Sanctus Unus" or rather "Ieoua Sanctus Unus" is actually an anagram of the Latin form of Newton's name: Isaacus Neuutonus.

5. "Quicksilver" or "mercury" was not only the name of the corresponding chemical element, but could also refer quite generally to substances with special qualities.

6. "Praxis," Ms. 420, The Babson College Grace K. Babson Collection of the Works of Sir Isaac Newton, Huntington Library, San Marino, California, USA.

7. The story that the fire was caused by Newton's dog "Diamond" knocking over a candle is definitely a later invention, however. Newton didn't have a dog. Nor did he have a cat, which is why the widespread story that he was the inventor of the cat flap is also nonsense (Newton is even supposed to have been stupid enough to make a large flap for his cat and an extra-small one for its young).

8. *Pantokrator*, from the Greek for "ruler of the world," is often used in the Greek Septuagint Bible and the New Testament as a synonym for "God."

9. M-J Liu, C-H Xiong, L Xiong, and X-L Huang, "Biomechanical Characteristics of Hand Coordination in Grasping Activities of Daily Living," *PLOS ONE* 11, 1 (2016): e0146193. doi:10.1371/journal.pone.0146193.

10. From "Newton and Alchemy" by Richard Westfall in *Occult & Scientific Mentalities in the Renaissance*, edited by Brian Vickers (Cambridge: Cambridge University Press, 1984), p. 316.

Chapter 7. Rivalry beyond Death: Newton and His Intrigues

1. Diogenes Laertius, DL IX 72.

2. Today, we know that the solution of Zeno's paradoxes is to be found precisely in this new understanding of infinity. The paradoxes only

vanished when, thanks to Leibniz and Newton, it was possible to deal mathematically and correctly with infinite additions, such as in the case of the race between Achilles and the tortoise. An infinite series need not necessarily be infinitely large, as Zeno thought it did.

3. But I certainly did move—my sweaty running gear is proof of that.

4. Erhard Weigel was not only an outstanding teacher at the university in my hometown of Jena—he was also an astronomer. While he didn't discover anything of note, he did think constantly about how best to pass on knowledge in a sensible and understandable way. When I was looking for a title for the internet blog I was planning to write, back in 2008, I therefore based my thinking on Weigel, since it was also my goal to make astronomy and science as clearly understandable as possible. One of Weigel's many inventions is a small instrument that simplifies orientation in the sky and with which you can show people where any stars can be found. This "simple star-pointer" or "Astrodicticum Simplex" became the name of my blog, in which I still try today to live up to Weigel's example (http://scienceblogs.de/astrodicticum-simplex/).

5. And today, the University of Leipzig is certainly terribly annoyed that they let such a prominent alumnus slip through their fingers.

6. See Eike Christian Hirsch, *Der berühmte Herr Leibniz* (Munich: C. H. Beck, 2017).

7. Before that, however, he sought contact with alchemists in Nuremberg, but unlike Newton, he had no particular enthusiasm for this. In order to be accepted by them, he had to prove himself to be worthy in his knowledge. He achieved this by simply writing a long text full of invented alchemistic nonsense, and nobody seemed to realize what he had done. He later wrote, in 1691, about his experiences: "Nuremberg first initiated me into chemical studies, and I do not regret having learned in my youth what makes me cautious as a man . . . I saw how . . . those well-known to me were shipwrecked, though they believed themselves to be sailing with the favorable winds of their alchemistic dreams."

8. A triangular number is a number which represents the sum of all numbers from 1 up to a certain upper limit. The first triangular number is 1. The second is then $1 + 2 = 3$, the next is $1 + 2 + 3 = 6$, followed by $1 + 2 + 3 + 4 = 10$, and so on. Leibniz was therefore asked to calculate the sum of $1 + 1/3 + 1/6 + 1/10 + \ldots$.

9. Naturally, Flamsteed immediately took advantage of the situation and sent a long list of Newton's mistakes and offenses to Leibniz.

10. "An Account of the Book Entitled *Commercium Epistolicum*," *Philosophical Transactions* 29 (1714/15).

11. For example, the integral symbol $\int$, which Leibniz first used in a paper dated October 29, 1675.

12. Leibniz was not only a diplomat, lawyer, and mathematician. He was a historian, and practiced geology, logic, linguistics, philology, and biology. He drew plans for an underwater boat, improved mining techniques, explored how to measure fever, and investigated all manner of other topics. Above all, though, he was also one of the most important philosophers of the modern era. It would go far beyond the limits of this book to present his philosophical works. But he was in any case superior to Newton in this field.

13. A dramatic case is described in the book *Plastic Fantastic* by Eugenie Reich, which looks at the faked research results of the German physicist Jan Hendrik Schön, who nevertheless managed to have thirteen papers published in highly respected journals before his misdemeanors were noticed.

14. The pressure to publish more and more quickly is accompanied by a whole new set of problems, of course, but that is a different story altogether.

INDEX